KT-406-430

Cambridge IGCSE™

Co-ordinated Sciences Chemistry

STUDENT'S BOOK

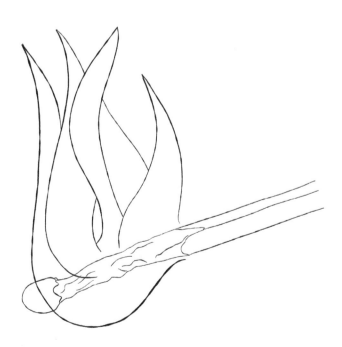

Chris Sunley and Sam Goodman

William Collins' dream of knowledge for all began with the publication of his first book in 1819. A self-educated mill worker, he not only enriched millions of lives, but also founded a flourishing publishing house. Today, staying true to this spirit, Collins books are packed with inspiration, innovation and practical expertise. They place you at the centre of a world of possibility and give you exactly what you need to explore it.

Collins. Freedom to teach.
Published by Collins
An imprint of HarperCollins*Publishers*
The News Building,
1 London Bridge Street,
London, SE1 9GF, UK

Harper*CollinsPublishers*
Macken House, 39/40 Mayor Street Upper,
Dublin 1, D01 C9W8, Ireland

Browse the complete Collins catalogue at
collins.co.uk

© HarperCollins*Publishers* Limited 2023

10 9 8 7 6 5 4 3 2 1

ISBN 978-0-00-854594-9

All rights reserved. No part of this publication may be reproduced, stored in a retrieval system, or transmitted in any form by any means, electronic, mechanical, photocopying, recording or otherwise, without the prior written permission of the Publisher or a licence permitting restricted copying in the United Kingdom issued by the Copyright Licensing Agency Ltd, 5th Floor, Shackleton House, 4 Battle Bridge Lane, London SE1 2HX.

British Library Cataloguing-in-Publication Data
A catalogue record for this publication is available from the British Library.

Authors: Chris Sunley and Sam Goodman
Publisher: Elaine Higgleton
Product manager: Joanna Ramsay
Content editors: Daniela Mora Chavarría, Renée Lewis
Development editor: Gillian Lindsey
Proof reader and answer checker: Just Content Ltd
Cover designer: Gordon MacGilp
Cover illustration: Ann Paganuzzi
Typesetter: Jouve India Private Limited
Production controller: Lyndsey Rogers
Printed and Bound in the UK using 100% Renewable
Electricity at CPI Group (UK) Ltd

This book is produced from independently certified FSC™ paper
to ensure responsible forest management.

For more information visit: www.harpercollins.co.uk/green

Endorsement indicates that a resource has passed Cambridge International's rigorous quality-assurance process and is suitable to support the delivery of a Cambridge International syllabus. However, endorsed resources are not the only suitable materials available to support teaching and learning, and are not essential to be used to achieve the qualification. Resource lists found on the Cambridge International website will include this resource and other endorsed resources.

Any example answers to questions taken from past question papers, practice questions, accompanying marks and mark schemes included in this resource have been written by the authors and are for guidance only. They do not replicate examination papers. In examinations the way marks are awarded may be different. Any references to assessment and/or assessment preparation are the publisher's interpretation of the syllabus requirements. Examiners will not use endorsed resources as a source of material for any assessment set by Cambridge International.

While the publishers have made every attempt to ensure that advice on the qualification and its assessment is accurate, the official syllabus, specimen assessment materials and any associated assessment guidance materials produced by the awarding body are the only authoritative source of information and should always be referred to for definitive guidance. Cambridge International recommends that teachers consider using a range of teaching and learning resources based on their own professional judgement of their students' needs.

Cambridge International has not paid for the production of this resource, nor does Cambridge International receive any royalties from its sale. For more information about the endorsement process, please visit www.cambridgeinternational.org/endorsed-resources

Cambridge International copyright material in this publication is reproduced under licence and remains the intellectual property of Cambridge Assessment International Education.

Third-party websites and resources referred to in this publication have not been endorsed by Cambridge Assessment International Education.

The publishers gratefully acknowledge the permission granted to reproduce the copyright material in this book. Every effort has been made to trace copyright holders and to obtain their permission for the use of copyright material. The publishers will gladly receive any information enabling them to rectify any error or omission at the first opportunity.

Chemistry

Getting the best from the book

Welcome to Collins *Cambridge IGCSE™ Co-ordinated Chemistry.*

This textbook has been designed to help you to succeed in this strand of your Cambridge IGCSE Co-ordinated Sciences syllabus study.

Just as there are 12 Sections in this strand of the syllabus, there are 12 sections in the textbook. Each section in the textbook covers the essential knowledge and skills you need. The textbook also has useful features which have been designed to help you understand all the aspects of Chemistry you will need to know for this syllabus.

SAFETY IN THE SCIENCE LESSON

This book is a textbook, not a laboratory or practical manual. As such, you should not interpret any information in this book that relates to practical work as including comprehensive safety instructions. Your teachers will provide full guidance for practical work and cover rules that are specific to your school.

A brief introduction gives context to the science covered in the section.

Starting points will help you to revise previous learning and see what you already know about the ideas in the section.

Section contents shows the chemistry topics of the syllabus covered in the section

This section starts with a recap on the differences between compounds and mixtures before introducing elements and their atomic structure. You will see why different elements have different properties and that the Periodic Table is a very useful way of ordering the elements. You will then consider the different ways atoms can join together to form compounds and how the method of combination will determine the properties of the compound formed. Finally you will study in more depth the bonding that is responsible for the characteristic properties of metals.

Starting points

1. What is an atom?

2. Can you name the particles that are found in atoms?

3. Can you explain the difference between a compound and a mixture?

4. Diamond and graphite are both covalent structures but have very different properties. Can you name some of the uses of diamond and graphite?

SYLLABUS SECTIONS COVERED

2.1 Elements, compounds and mixtures

2.2 Atomic structure and the Periodic Table

2.3 Isotopes

2.4 Ions and ionic bonds

2.5 Simple molecules and covalent bonds

2.6 Giant covalent structures

2.7 Metallic bonding

2
Atoms, elements and compounds

Knowledge check reminds you of the ideas you should have already encountered in previous work before starting the topic.

Learning objectives show you what the syllabus requires in the topic.

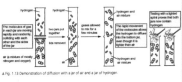
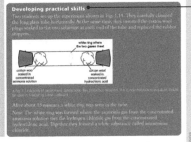

Examples of investigations are included with questions matched to the practical skills you will need to learn. It is not expected you will need to perform or learn all the methods of the example investigations as they extend beyond the requirements of the syllabus.

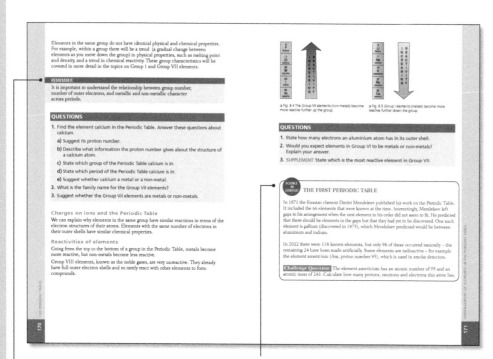

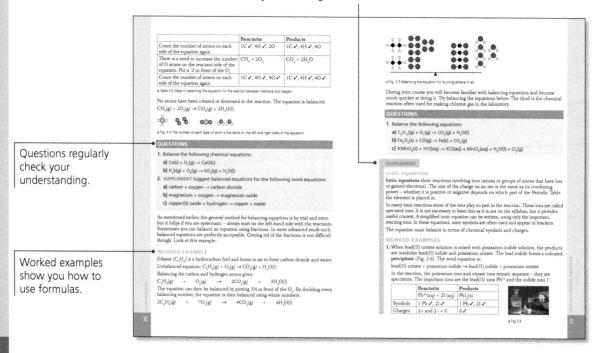

Remember boxes provide tips and guidance to help you during your course.

Science in context boxes put the ideas you are learning into real-life context. The content in these boxes is beyond the requirements of the syllabus. However, they do provide interesting examples of scientific application that are designed to enhance your understanding and a challenge question to encourage you to think more deeply beyond the syllabus content.

Clearly differentiated Supplement material is only relevant for students studying the Extended syllabus content and so helps you focus your learning.

Questions regularly check your understanding.

Worked examples show you how to use formulas.

End of topic questions allow you to apply the knowledge and understanding you have learned in the topic to answer the questions.

Key terms for the topic are defined in the glossary

A full checklist of all the information you need to cover the complete chemistry content covered in each topic.

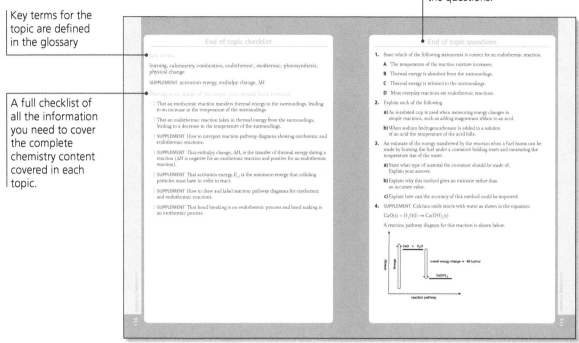

Regular practice questions provide a checkpoint for your learning and help you prepare for your exams.

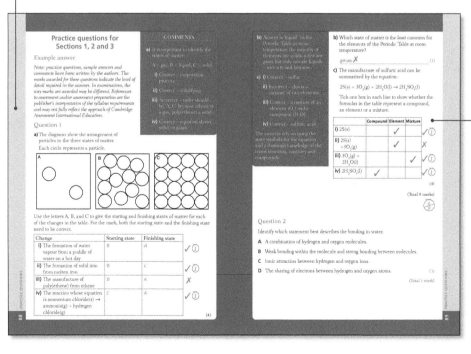

Example answers and comments have been written by the authors to illustrate a possible response and show where it answers the question with the level of detail required or how it could be improved. Practice questions and sample answers have been written by the authors. In examinations, the way marks are awarded may be different.

GETTING THE BEST FROM THE BOOK

7

This section will provide a recap of the information you will probably have encountered in your previous work. It provides a foundation for many of the other sections in this book. As well as basic information on the states of matter, importantly it introduces the kinetic particle theory which will feature in many of the remaining sections.

Starting points

1. Can you name the three common states of matter?

2. How many of the processes involved in changing one state of matter into another could you list?

3. Do you know what diffusion is?

SYLLABUS SECTIONS COVERED

1.1 Solids, liquids and gases

1.2 Diffusion

1
States of matter

Δ On the edge of the Arctic Circle.

△ Fig. 1.1 Water in all its states of matter.

Solids, liquids and gases

INTRODUCTION

Nearly all substances may be classified as solid, liquid or gas – the states of matter. In science these states are often shown in shorthand as (s), (l) and (g) after the formula or symbol. The differences between solids, liquids and gases can be explained using the idea that all substances are made up of extremely tiny particles. The particles in these three states are arranged differently and have different types of movement and different energies. In many cases, matter changes into different states quite easily. The names of many of these processes are in everyday use, such as melting and condensing. Using simple models of the particles in solids, liquids and gases can help to explain what happens when a substance changes state.

KNOWLEDGE CHECK

✓ Substances are classified as solid, liquid or gas.
✓ Solids, liquids and gases have different properties.
✓ All substances are made up of particles.

LEARNING OBJECTIVES

✓ State the distinguishing properties of solids, liquids and gases.
✓ Describe the structures of solids, liquids and gases in terms of particle separation, arrangement and motion.
✓ Describe changes of state in terms of melting, boiling, evaporating, freezing and condensing.
✓ Describe the effects of temperature and pressure on the volume of a gas.
✓ SUPPLEMENT Explain changes of state in terms of kinetic particle theory, including the interpretation of heating and cooling curves.
✓ SUPPLEMENT Explain, in terms of kinetic particle theory, the effects of temperature and pressure on the volume of a gas.
✓ Describe and explain diffusion in terms of kinetic particle theory.
✓ SUPPLEMENT Describe and explain the effect of relative molecular mass on the rate of diffusion of gases.

△ Fig. 1.2 Water covers nearly four-fifths of the Earth's surface. In this photo you can see that all three states of matter can exist together: solid water (the ice) is floating in liquid water (the ocean), and the surrounding air contains water vapour (clouds).

STRUCTURE AND PROPERTIES OF SOLIDS, LIQUIDS AND GASES

There are three states of matter shown in shorthand as (s), (l) and (g) after the formula or symbol of an element or compound. The symbols (s), (l) and (g) are called **state symbols**. The three states of matter each have different properties, depending on how strongly the particles are held together.

- **Solids** have a fixed volume and shape.
- **Liquids** have a fixed volume but no definite shape. They take up the shape of the container in which they are held.
- **Gases** have no fixed volume or shape. They spread out to fill whatever container or space they are in.

Substances don't always exist in the same state; depending on the physical conditions, they change from one state to another (interconvert).

Some substances can exist in all three states in the natural world.

A good example of this is water, as shown in Fig. 1.2.

QUESTIONS

1. Give the state symbol for a liquid.
2. Which is the only state of matter that has a fixed shape?
3. Describe the ways in which fine sand behaves like a liquid.

Why do solids, liquids and gases have different properties?

The behaviour of solids, liquids and gases can be explained if we think of all matter as being made up of very small particles that are in constant motion. This idea has been summarised in the **particle theory**.

In solids, the particles are held tightly together in a fixed position, so solids have a definite shape. However, the particles are vibrating about their fixed positions.

In liquids, the particles are held tightly together but can move around move around. Liquids have no definite shape and will take on the shape of the container they are in.

In gases, the particles are further apart and are constantly moving. Gas particles can spread apart to fill the container they are in.

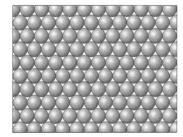
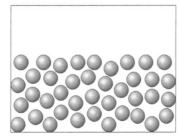
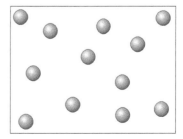

Δ Fig. 1.3 Particles in a solid.   Δ Fig. 1.4 Particles in a liquid.   Δ Fig. 1.5 Particles in a gas.

Changes of state

The temperatures at which one state changes to another have specific names:

Name of the process	Change of state
Melting	Solid to liquid
Boiling	Liquid to gas
Freezing	Liquid to solid
Evaporating	Liquid to gas
Condensing	Gas to liquid

Δ Table 1.1 Changes of state.

The particles in a liquid can move around. They have different energies, so some are moving faster than others. The faster particles have enough energy to escape from the surface of the liquid and they change into the gas state (also called **vapour** particles). This process is **evaporation**. The rate of evaporation increases with increasing temperature because heating gives more particles the energy to be able to escape from the surface.

Fig. 1.6 summarises the changes in states of matter. Note that **melting** and **freezing** happen at the same temperature – as do **boiling** and **condensing**.

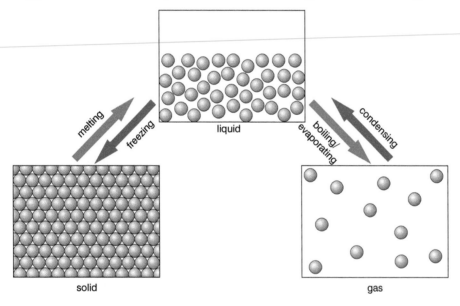

Δ Fig. 1.6 Changes in states of matter. Note that melting and freezing happen at the same temperature – as do boiling and condensing.

How do substances change from one state to another?

The changes of state can be explained using the kinetic particle theory.

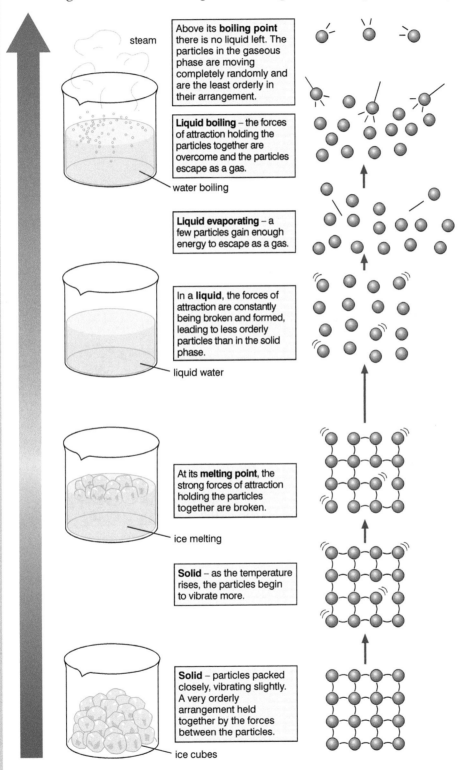

Above its **boiling point** there is no liquid left. The particles in the gaseous phase are moving completely randomly and are the least orderly in their arrangement.

Liquid boiling – the forces of attraction holding the particles together are overcome and the particles escape as a gas.

water boiling

Liquid evaporating – a few particles gain enough energy to escape as a gas.

In a **liquid**, the forces of attraction are constantly being broken and formed, leading to less orderly particles than in the solid phase.

liquid water

At its **melting point**, the strong forces of attraction holding the particles together are broken.

ice melting

Solid – as the temperature rises, the particles begin to vibrate more.

Solid – particles packed closely, vibrating slightly. A very orderly arrangement held together by the forces between the particles.

ice cubes

steam

Δ Fig. 1.7 Explaining changes of state in terms of kinetic particle theory.

To change solids into liquids and then into gases, thermal energy must be put in. Heating provides the particles with enough energy to overcome the forces holding them together.

To change gases into liquids and then into solids involves cooling, so removing thermal energy. This makes the particles come closer together as the substance changes from gas to liquid and the forces of attraction between the particles increase as the liquid becomes a solid.

THE STATES OF MATTER

There are three states of matter – or are there? To complicate this simple idea, some substances show the properties of two different states of matter. Some examples are given below.

Liquid crystals

Liquid crystals are commonly used in displays in computers and televisions. Within particular temperature ranges the particles of the liquid crystal can flow like a liquid, but remain arranged in a pattern in which the particles cannot rotate.

Superfluids

When some liquids are cooled to very low temperatures they form a second liquid state, described as a *superfluid* state. Liquid helium at just above the temperature of absolute zero has infinite fluidity and will 'climb out' of its container when left undisturbed – at this temperature the liquid has zero viscosity. (You may like to look up 'fluidity' and 'viscosity'.)

△ Fig. 1.8 An LCD (liquid crystal display) television.

Plasma

Plasmas, or ionised gases, can exist at temperatures of several thousand degrees Celsius. An example of a plasma is the charged air produced by lightning. Stars like our Sun also produce plasma. Like a gas, a plasma does not have a definite shape or volume but the strong forces between its particles give it unusual properties, such as conducting electricity. Because of this combination of properties, plasma is sometimes called the fourth state of matter.

Challenge Question: 'Fine sand behaves like a liquid crystal'. Suggest a common property of these two materials which supports this statement.

Changing the volume of a gas

Everyday examples illustrate these changes. For example, the size of an inflated balloon increases on a hot day but it decreases if the balloon is squeezed. Increasing temperature will increase the **volume** of a gas. The particles of the gas move more quickly and take up more space. Increasing **pressure** has the opposite effect – the particles move closer together and so the volume of the gas decreases.

SUPPLEMENT

The **pressure** exerted by a gas is caused by the gas particles bombarding the sides of the container the gas is in. If a gas is compressed into a smaller volume, the number of gas particles hitting the sides of the container every second will increase (there is less space for them to move in). The increase in the number of collisions per second causes the increase in pressure. If the temperature of a gas is increased, the gas particles have more energy and will move faster. This will cause an increase in volume. If the gas is in a container then there will be more collisions with the sides of the container each second and so the gas will exert a greater pressure.

△ Fig. 1.9 A heating curve.

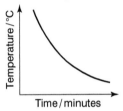

△ Fig. 1.10 A cooling curve.

The graphs show a heating curve and a cooling curve for water, first being heated in a kettle and then cooling after the heating has stopped. The heating curve shows the temperature increasing quickly (the curve has a higher gradient) and then beginning to level off (the curve has a lower gradient) as the water gets closer to its boiling point. The particles increase in movement more quickly as the water is heated but the increase in particle movement decreases as the water approaches boiling point. The cooling curve shows the temperature falling, quickly at first and then more slowly as the water temperature nears the room temperature. When the heating is stopped water particles lose kinetic energy more quickly initially and then, as the cooling process continues, the loss of kinetic energy is more gradual as the water gets closer to room temperature.

QUESTIONS

1. Describe the type of movement the particles in a solid have.
2. In which state are the particles held together more strongly: in solid water, liquid water or water vapour?
3. What is the name of the process that occurs when the faster-moving particles in a liquid escape from its surface?
4. What is the name given to the process in which a solid changes into a liquid?
5. SUPPLEMENT Explain in terms of kinetic particle theory why increasing pressure will reduce the volume of a gas.

6. Which two of these changes will increase the volume of a gas?

    A increase temperature

    B increase pressure

    C reduce pressure

    D reduce temperature

DIFFUSION

Scientists have confidence in the kinetic particle theory because of the evidence from simple experiments.

The random mixing and moving of particles in liquids and gases is known as **diffusion**. The examples given below show the effects of diffusion.

Dissolving crystals in water

Fig. 1.11 shows purple crystals of potassium manganate(VII) dissolving in water.

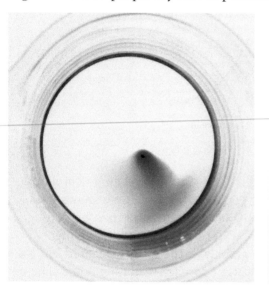

△ Fig. 1.11 Crystals of potassium manganate(VII) dissolving in water. The picture on the left shows the water immediately after the crystal was added; the picture on the right shows the water 1 hour later.

There are no water currents, so only the kinetic particle theory can explain this. The particles of the crystal gradually move into the water and mix with the water particles.

Mixing gases

These photos show a jar of air and a jar of bromine gas. Bromine gas is red-brown and heavier than air. The jar of air has been placed on top of the jar of bromine and the lids removed so the gases can mix (left-hand part of Fig.1.12).

After about 24 hours the bromine gas and the air have spread throughout both jars. Kinetic particle theory says that the particles of bromine gas can move around randomly so that they can fill both gas jars. This also occurs with hydrogen and air (Fig. 1.13).

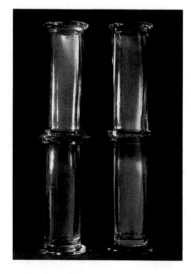

△ Fig. 1.12 Diffusion of bromine.

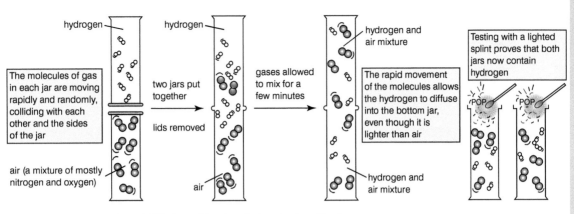

The molecules of gas in each jar are moving rapidly and randomly, colliding with each other and the sides of the jar

air (a mixture of mostly nitrogen and oxygen)

hydrogen

two jars put together

lids removed

air

hydrogen

gases allowed to mix for a few minutes

hydrogen and air mixture

The rapid movement of the molecules allows the hydrogen to diffuse into the bottom jar, even though it is lighter than air

hydrogen and air mixture

Testing with a lighted splint proves that both jars now contain hydrogen

POP   POP

△ Fig. 1.13 Demonstration of diffusion with a jar of air and a jar of hydrogen.

The molecules of different gases do not all move at the same speed at room temperature. The rate at which gases diffuse depends on their relative molecular mass – the larger their relative molecular mass, the lower their rate of diffusion. For example hydrogen, which has the lowest relative molecular mass of any gas, will diffuse much more rapidly than carbon dioxide which has a relative molecular mass 22 times that of hydrogen.

Developing practical skills

Two students set up the experiment shown in Fig. 1.14. They carefully clamped the long glass tube horizontally. At the same time, they inserted the cotton wool plugs soaked in the two solutions at each end of the tube and replaced the rubber stoppers.

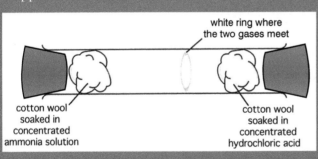

white ring where the two gases meet

cotton wool soaked in concentrated ammonia solution

cotton wool soaked in concentrated hydrochloric acid

△ Fig. 1.14 Results of experiment. Safety note: Eye protection required. The concentrated acid and alkali should be used in a working fume cupboard.

After about 15 minutes a white ring was seen in the tube.

Note: The white ring was formed where the ammonia gas from the concentrated ammonia solution met the hydrogen chloride gas from the concentrated hydrochloric acid. Together they formed a white substance called ammonium chloride.

Using and organising techniques, apparatus and materials

The concentrated ammonia solution is corrosive – it burns and is dangerous to the eyes. Concentrated hydrochloric acid is corrosive – it burns and its vapour irritates the lungs.

1. How should the cotton wool plugs have been handled when putting them into the tube?

2. Suggest other safety precaution(s) the student should have used.

Observing, measuring and recording

3. Deduce which gas moved furthest in the 15 minutes before the ring formed.

4. Determine approximately how much further this gas travelled compared to the other gas.

Handling experimental observations and data

5. The rate of diffusion of a gas depends on the mass of its particles. State what conclusion you can make about the relative masses of the two gases in this experiment.

QUESTIONS

1. Suggest a definition for *diffusion*.

2. Explain how the purple colour of the potassium manganate(VII) shown in Fig. 1.11 spreads through the water.

3. A bottle of perfume is broken at one end of a room. Explain why the perfume can soon be smelled all over the room.

4. SUPPLEMENT Use the idea of relative molecular mass to explain why hydrogen diffuses more quickly than oxygen.

End of topic checklist

Key terms

boiling, boiling point, condensing, diffusion, evaporating, freezing, gas, particle theory, liquid, melting, melting point, pressure, solid, state symbols, vapour

SUPPLEMENT kinetic particle theory, relative molecular mass (M_r)

During your study of this topic you should have learned:

○ About the different properties of solids, liquids and gases.

○ How to describe the structure of solids, liquids and gases in terms of particle separation, arrangement and types of motion.

○ How to describe changes of state in terms of melting, boiling, evaporating, freezing and condensing.

○ How to describe the effects of pressure and temperature on the volume of a gas.

○ How to describe and explain diffusion in terms of kinetic particle theory.

○ **SUPPLEMENT** How to explain changes of state in terms of the kinetic particle theory, including the interpretation of heating and cooling curves.

○ **SUPPLEMENT** How to explain, in terms of kinetic particle theory, the effects of temperature and pressure on the volume of a gas.

○ **SUPPLEMENT** How to describe and explain that the rate of diffusion of gases depends on their relative molecular masses.

End of topic questions

1. Identify in which of the three states of matter the particles move fastest.

2. Describe the arrangement and movement of the particles in a liquid.

3. Identify in which state of matter the particles just vibrate about a fixed point.

4. Sodium (melting point 98 °C) and aluminium (melting point 660 °C) are both solids at room temperature. From their melting points, state what you can conclude about the forces of attraction between the particles in the two metals.

5. Identify one of the processes shown as an example of evaporation.

    A $Fe(s) \rightarrow Fe(l)$

    B $H_2O(l) \rightarrow H_2O(g)$

    C $H_2O(g) \rightarrow H_2O(l)$

    D $H_2O(l) \rightarrow H_2O(s)$

6. Ethanol liquid turns into ethanol vapour at 78 °C. State the name of this temperature.

7. Explain how water in the Earth's polar regions can produce water vapour even when the temperature is very low.

8. A student wrote in her exercise book, 'The particle arrangement in a liquid is more like the arrangement in a solid than in a gas.' Do you agree with this statement? Justify your reasoning.

9. State the word that describes the rapid mixing and moving of particles in gases.

10. Look at Fig. 1.12 showing gas jars of air and bromine. Explain why bromine gas fills the top gas jar even though it is denser than air.

11. **SUPPLEMENT** Use kinetic particle theory to explain:

    a) the effect of temperature on the volume of a gas.

    b) the effect of pressure on the volume of a gas.

12. SUPPLEMENT The relative molecular masses of some gases are shown in the table.

Predict which gas would diffuse at the greatest rate. Explain your answer.

Gas	Relative molecular mass
Oxygen	32
Nitrogen	28
Chlorine	71

This section starts with a recap on the differences between compounds and mixtures before introducing elements and their atomic structure. You will see why different elements have different properties and that the Periodic Table is a very useful way of ordering the elements. You will then consider the different ways atoms can join together to form compounds and how the method of combination will determine the properties of the compound formed. Finally you will study in more depth the bonding that is responsible for the characteristic properties of metals.

Starting points

1. What is an atom?
2. Can you name the particles that are found in atoms?
3. Can you explain the difference between a compound and a mixture?
4. Diamond and graphite are both covalent structures but have very different properties. Can you name some of the uses of diamond and graphite?

SYLLABUS SECTIONS COVERED

2.1 Elements, compounds and mixtures

2.2 Atomic structure and the Periodic Table

2.3 Isotopes

2.4 Ions and ionic bonds

2.5 Simple molecules and covalent bonds

2.6 Giant covalent structures

2.7 Metallic bonding

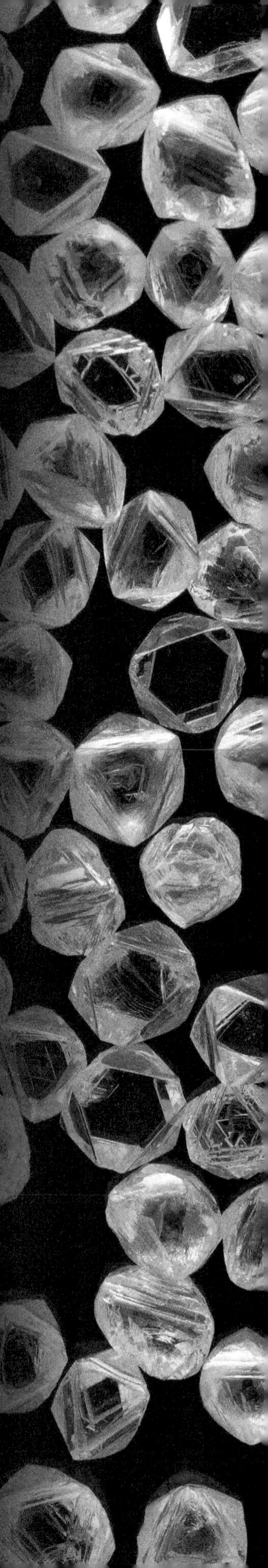

2
Atoms, elements and compounds

△ Some crystals are found in the natural world.

△ Fig. 2.1 Non metals: from left: silicon, chlorine, sulfur.

Materials, atomic structure and the Periodic Table

INTRODUCTION

This topic is about the structure, or the makeup, of all substances. Some substances exist in nature as elements, others as compounds that are formed when elements combine chemically, and others as mixtures. The topic starts by considering the structure of the atoms that make up elements. It shows how the arrangement of elements in the Periodic Table is determined by the structure of their atoms. The link between electronic structures and metals and non-metals is introduced. The following topics look in more detail at how atoms combine together to form ions and molecules, and the structure of metals.

KNOWLEDGE CHECK

✓ There are three states of matter and the kinetic particle theory explains what happens when one state is converted into another.
✓ Diffusion experiments provide evidence for the existence of particles.
✓ Compounds are formed when elements combine together chemically.

LEARNING OBJECTIVES

✓ Describe the differences between elements, compounds and mixtures.
✓ Describe the structure of the atom as a central nucleus containing neutrons and protons surrounded by electrons in shells.
✓ State the relative charges and relative masses of a proton, a neutron and an electron.
✓ Define proton number / atomic number as the number of protons in the nucleus of an atom.
✓ Define mass number / nucleon number as the total number of protons and neutrons in the nucleus of an atom.
✓ Determine the electronic configuration of elements and their ions with proton number 1 to 20, e.g. 2,8,3.
✓ State that: Group VIII noble gases have a full outer shell; the number of outer shell electrons is equal to the group number in Groups I to VII; the number of occupied electron shells is equal to the period number.
✓ Define isotopes as different atoms of the same element that have the same number of protons but different numbers of neutrons.
✓ Interpret and use symbols for atoms, e.g. $^{12}_{6}C$, and ions, e.g. $^{35}_{17}Cl^{-}$
✓ **SUPPLEMENT** State that isotopes of the same element have the same chemical properties because they have the same number of electrons and therefore the same electronic configuration.

ELEMENTS, COMPOUNDS AND MIXTURES

All matter can be classified into the three categories of elements, mixtures and compounds.

An **element** is the smallest part of a substance that can exist on its own. When two or more elements combine together a **compound** is formed. In a compound the **atoms** or elements are chemically combined together.

A **mixture** contains more than one substance (elements or compounds). In a mixture, the individual substances can be separated by simple means. This is because the substances in a mixture have not combined chemically.

ATOMIC STRUCTURE AND THE PERIODIC TABLE

In 1808 the British chemist John Dalton published a book outlining his theory of atoms. These were the main points of his theory:

- All matter is made of small, indivisible spheres called atoms.
- All the atoms of a given element are identical and have the same mass.
- The atoms of different elements have different masses.
- Chemical compounds are formed when different elements join together.

Since 1808, atomic theory has developed considerably and yet many of Dalton's ideas are still correct. Modern theory is built on an understanding of the particles that make up atoms – the so-called sub-atomic particles.

Sub-atomic particles

The smallest amount of an element that still behaves like that element is an atom. Each element has its own unique type of atom. Atoms are made up of smaller, sub-atomic particles. The three main sub-atomic particles are **protons**, **neutrons** and **electrons**.

These particles are very small and have very little mass. However, it is possible to compare their masses to the mass of a proton. Their charges may also be compared in a similar way. These numbers are called the relative mass and the relative charge. The proton and neutron have the same mass, and the proton and electron have equal but opposite charges. As an atom has equal numbers of protons and electrons atoms are neutral.

Sub-atomic particle	Relative mass	Relative charge
Proton	1	+1
Neutron	1	0
Electron	about $\frac{1}{2000}$	−1

△ Table 2.1 Relative masses and charges of sub-atomic particles.

Protons and neutrons are found in the centre of the atom in a cluster called the **nucleus**. The electrons form a series of **shells** around the nucleus. The shells are different distances from the nucleus.

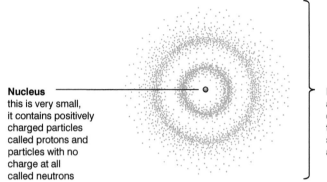

Nucleus
this is very small,
it contains positively
charged particles
called protons and
particles with no
charge at all
called neutrons

Electrons
are negatively
charged particles
that form a
series of 'shells'
around the nucleus

Δ Fig. 2.2 Structure of an atom.

Proton number and mass number

In order to describe the numbers of protons, neutrons and electrons in an atom, scientists use two numbers. These are called the **proton number** (or **atomic number**) and the **mass number** (or **nucleon number**). The proton number, as you might expect, describes the number of protons in the nucleus of an atom. The mass number describes the number of particles in the nucleus of the atom – that is, the total number of protons and neutrons.

Proton numbers are used to arrange the elements in the Periodic Table. The atomic structures of the first ten elements in the Periodic Table are shown in Table 2.2.

Hydrogen is the only atom that has no neutrons.

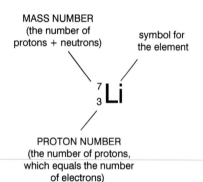

MASS NUMBER
(the number of
protons + neutrons)

symbol for
the element

$^{7}_{3}\text{Li}$

PROTON NUMBER
(the number of protons,
which equals the number
of electrons)

Δ Fig. 2.3 The chemical symbol for lithium showing the mass number and proton number.

Element	Proton number	Mass number	Number of protons	Number of neutrons	Number of electrons
Hydrogen	1	1	1	0	1
Helium	2	4	2	2	2
Lithium	3	7	3	4	3
Beryllium	4	9	4	5	4
Boron	5	11	5	6	5
Carbon	6	12	6	6	6
Nitrogen	7	14	7	7	7
Oxygen	8	16	8	8	8
Fluorine	9	19	9	10	9
Neon	10	20	10	10	10

Δ Table 2.2 Atomic structures of the first ten elements.

A **Periodic Table** of elements is shown on page 318 of this book and you will recognise the names of some of the more common elements that you know about, for example, carbon (C), oxygen (O), aluminium (Al) and iron (Fe).

1. Which sub-atomic particle has the smallest relative mass?
2. Why do atoms have the same number of protons as electrons?
3. An aluminium atom can be represented as $^{27}_{13}$Al.

    a) What is aluminium's mass number?

    b) Calculate how many neutrons this atom of aluminium has.

ISOTOPES

Atoms of the same element with the same number of protons and electrons but different numbers of neutrons are called **isotopes**. For example, there are two isotopes of chlorine:

Symbol	Number of protons	Number of neutrons
$^{35}_{17}$Cl	17	18
$^{37}_{17}$Cl	17	20

Δ Table 2.3 Isotopes of chlorine.

SUPPLEMENT

The chemical properties of an element depend on the number of electrons in the outer electron shell. As isotopes of the same element have the same electron configuration they have the same chemical properties.

SCIENCE IN CONTEXT ISOTOPES

Some isotopes have different physical properties. Some isotopes are radioactive. They emit radioactivity from the nucleus and decay – that is, if the radioactivity is alpha particles or beta particles, they change into other atoms with different numbers of protons and/or neutrons.

Isotopes that do not decay and do not emit radiation are classed as non-radioactive isotopes.

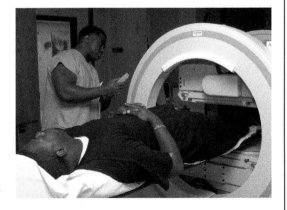

Δ Fig. 2.4 A gamma camera is an example of radioactive isotopes used in medicine.

Radioactive isotopes emit radioactivity of one of the following types:

Alpha particles (α) are helium nuclei; they contain two protons and two neutrons.

Beta particles (β) are fast-moving electrons.

Gamma rays (γ) are high-energy electromagnetic rays.

In medicine, radioactive isotopes are used to sterilise equipment and in radiotherapy to treat cancer tumours.

Industry uses radioactive isotopes as 'tracers' to detect leaks in pipes.

Challenge Question: Suggest how you could use a radioactive tracer to detect a leak in a sewage pipe.

SCIENCE IN CONTEXT

SUB-ATOMIC PARTICLES

Protons, neutrons and electrons are the particles from which atoms are made. However, in the past 20 years or so scientists have discovered a number of other sub-atomic particles: quarks, leptons, muons, neutrinos, bosons and gluons. The properties of some of these particles have become well known, but there is still much to learn about the others. Finding out about these, and possibly other sub-atomic particles, is one of the challenges of the 21st century.

△ Fig. 2.5 The Large Hadron Collider at CERN in Switzerland.

To study the smallest known particles, a particle accelerator has been built underground at CERN near Geneva, Switzerland. This giant instrument, called the Large Hadron Collider (LHC), has a circumference of 27 km. It attempts to recreate the conditions that existed just after the 'Big Bang' by colliding beams of particles at very high speed – only about 5 m/s slower than the speed of light. It promises to revolutionise scientific understanding of the nature of atoms. Who knows – school science in 10 or 20 years' time may be very different from your lessons today!

Challenge Question: Describe what you can find out about the 'Big Bang' theory.

QUESTIONS

1. What are *isotopes*?

2. SUPPLEMENT Explain why different isotopes of the same element have the same chemical properties.

ELECTRONIC CONFIGURATION
Arrangements of electrons in the atom

An atom's electrons are arranged in shells around the nucleus. These do not all contain the same number of electrons – the shell nearest to the nucleus can take only two electrons, whereas the next one out from the nucleus can take eight.

Electron shell	Maximum number of electrons
1	2
2	8
3	8

△ Table 2.4 Maximum number of electrons in a shell.

Oxygen has a proton number of 8, so it has 8 electrons. Of these, two are in the first shell and six are in the second shell. This arrangement is written 2, 6. A phosphorus atom with a proton number of 15 has 15 electrons, arranged 2, 8, 5. It is the electrons in the outer electron shell that are involved in chemical bonding, which you will learn more about in the next two topics.

Atom diagrams

The atomic structure of an atom can be shown simply in a diagram. The arrangement of electrons in shells around the nucleus of an atom can be shown simply in a diagram, using a dot for each electron.

The arrangement of electrons in an atom is called its **electronic configuration**. This can also be written as a sequence of

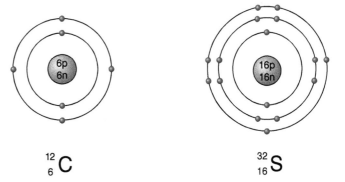

$$^{12}_{6}C \qquad\qquad ^{32}_{16}S$$

△ Fig. 2.6 Atom diagrams for carbon and sulfur showing the numbers of protons and neutrons and the electron arrangements.

numbers showing the number of electrons in each shell. For example, the electronic configuration of carbon is 2, 4 and the electronic configuration of sulfur is 2, 8, 6.

Electronic configuration: The first 20 elements of the Periodic Table

There are over 110 different elements. They are arranged in the Periodic Table in order of increasing proton number. They are then arranged in **periods** and **groups** according to their chemical and physical properties.

The chemical properties of elements depend on the arrangement of electrons in their atoms. The electronic configuration of the first 20 elements is shown in Table 2.5.

Element	Symbol	Proton number	Electron number	Electronic configuration
Hydrogen	H	1	1	1
Helium	He	2	2	2
Lithium	Li	3	3	2, 1

Element	Symbol	Proton number	Electron number	Electronic configuration
Beryllium	Be	4	4	2, 2
Boron	B	5	5	2, 3
Carbon	C	6	6	2, 4
Nitrogen	N	7	7	2, 5
Oxygen	O	8	8	2, 6
Fluorine	F	9	9	2, 7
Neon	Ne	10	10	2, 8
Sodium	Na	11	11	2, 8, 1
Magnesium	Mg	12	12	2, 8, 2
Aluminium	Al	13	13	2, 8, 3
Silicon	Si	14	14	2, 8, 4
Phosphorus	P	15	15	2, 8, 5
Sulfur	S	16	16	2, 8, 6
Chlorine	Cl	17	17	2, 8, 7
Argon	Ar	18	18	2, 8, 8
Potassium	K	19	19	2, 8, 8, 1
Calcium	Ca	20	20	2, 8, 8, 2

Δ Table 2.5 Electronic configuration of first 20 elements.

In the Periodic Table, the columns are called **groups** and the rows are called **periods**. In the Periodic Table (see page 318) lithium, sodium and potassium are placed on the left, and neon and argon are placed on the right. The proton number increases from lithium to neon, moving through a section, or **period**, of the Periodic Table.

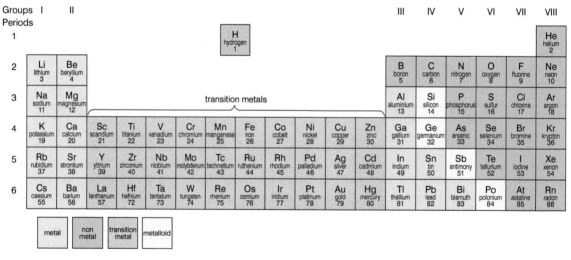

Δ Fig. 2.7 The Periodic Table.

Electronic configuration and group period numbers

The number of occupied electronic shells is equal to the period number. So, for example, sodium is in the 3rd period and potassium is in the 4th period.

For elements in Groups I to VII the number of outer shell electrons is equal to the group number. For example lithium, sodium and potassium each have one electron in the outermost shell, as shown in Table 2.5 and Figure 2.8. They are all Group I elements. Sodium and potassium both have one electron in the outermost shell – they are Group I elements. Carbon and silicon have 4 electrons in their outermost shell – they are Group IV elements.

Electronic configuration and chemical properties

Elements that have similar electronic configurations have similar chemical properties. **Metals** are usually found to the left and middle of the Periodic Table but **non-metals** are usually found on the right.

Lithium (2, 1), sodium (2, 8, 1) and potassium (2, 8, 8, 1) all have one electron in their outer shell. These are all highly reactive metals. They are called Group I elements.

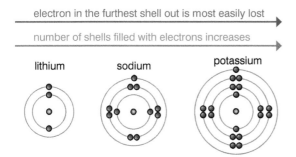

△ Fig. 2.8 Electronic configurations of lithium, sodium and potassium.

Fluorine (2, 7), chlorine (2, 8, 7), bromine (2, 8, 18, 7) and iodine (2, 8, 18, 18, 7) all have seven electrons in their outer shell. These elements are all highly reactive **non-metals**. They are called Group VII elements, or halogens.

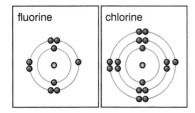

△ Fig. 2.9 Electronic configuration of fluorine and chlorine.

Similarly, all the elements in Group III of the Periodic Table have three electrons in their outer electron shell.

The elements helium (2), neon (2, 8) and argon (2, 8, 8) either have a full outer shell or have eight electrons in their outer shell. These elements are in group VIII and are called **noble gases**.

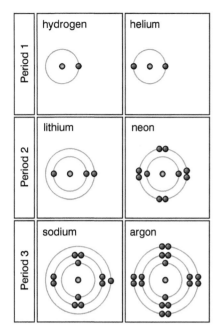

△ Fig. 2.10 Electronic configurations of helium, neon and argon.

△ Fig. 2.11 Neon lighting in Hong Kong.

QUESTIONS

1. a) How many electrons does magnesium have in its outer electron shell?

   b) Suggest which group of the Periodic Table magnesium is in.

2. Sketch atom diagrams for:

   a) aluminium

   b) calcium.

End of topic checklist

Key terms

Atom, compound, electron, electronic configuration, element, group, isotopes, metal, mixture, neutron, noble gases, non-metal, mass number (nucleon number), nucleus, period, Periodic Table, proton, proton number (atomic number), shell

During your study of this topic you should have learned:

○ How to describe the differences between elements, compounds and mixtures.

○ How to describe the structure of an atom as a central nucleus containing neutrons and protons surrounded by electrons in shells.

○ The relative charges and approximate relative masses of protons, neutrons and electrons.

○ How to define *proton number* (atomic number) and *mass number* (nucleon number).

○ How to describe the electronic configuration of elements with proton numbers 1 to 20.

○ That the arrangement of electrons in an atom can be worked out from the atom's position in the Periodic Table.

○ How to explain the build-up of electrons in 'shells' and the significance of the noble gas electronic structures, period number and the link between outer electrons and Group number.

○ How to define *isotopes*.

○ How to interpret and use symbols for atoms.

○ **SUPPLEMENT** That isotopes have the same chemical properties because they have the same number of electrons in their outer shell.

End of topic questions

1. Identify which of the following statements is true.

   A A proton has the same relative mass as an electron.

   B An electron has the same relative mass as a neutron.

   C A proton has the same relative mass as a neutron.

   D A proton and an electron have the same relative charge.

2. Explain the meaning of:

   a) *proton number* (atomic number)

   b) *mass number* (nucleon number).

3. Chlorine has two common isotopes, chlorine-35 and chlorine-37.

   a) What is an *isotope?*

   b) Calculate the numbers of protons, neutrons and electrons in each isotope.

4. Copy and complete the table.

Atom	Number of protons	Number of neutrons	Number of electrons	Electron configuration
$^{28}_{14}\text{Si}$				
$^{24}_{12}\text{Mg}$				
$^{32}_{16}\text{S}$				
$^{40}_{18}\text{Ar}$				

5. The table shows information about the structure of six particles (A–F).

Particle	Protons	Neutrons	Electrons
A	8	8	10
B	12	12	10
C	6	6	6
D	8	10	10
E	6	8	6
F	11	12	11

a) In each of questions **i)** to **v)**, identify one of the six particles A–F. Each letter may be used once, more than once or not at all.

Identify a particle that:

   i) has a mass number of 12

   ii) has the highest mass number

   iii) has no overall charge

   iv) has an overall positive charge

   v) is the same element as particle E.

b) Sketch an atom diagram for particle E.

6. Draw an atom diagram for:

   a) oxygen

   b) potassium.

7. For each of parts **a)** to **d)**, state whether the statement is TRUE or FALSE. There is a relationship between the group number of the first 20 elements in the Periodic Table and:

   a) the number of protons in an atom of the element

   b) the number of neutrons in an atom of the element

   c) the number of electrons in an atom of the element

   d) the number of electrons in the outer electron shell of the element.

8. **SUPPLEMENT** There are two common isotopes of chlorine, chlorine-35 and chlorine-37. Compare and contrast these two isotopes and explain why they have exactly the same chemical properties.

Ions and ionic bonds

△ Fig. 2.12 Sodium chloride is an example of an ionic compound.

INTRODUCTION

When the atoms of elements react and join together, they form compounds. When one of the reacting atoms is a metal, the compound formed is called an ionic compound. Ionic compounds do not contain molecules; instead they are made of particles called ions. Ionic compounds have similar physical properties, many of which are quite different from the properties of substances made up of atoms or molecules.

KNOWLEDGE CHECK

✓ Compounds are formed when the atoms of two or more elements combine together.
✓ Protons have a positive charge and are found in the nucleus of the atom.
✓ Electrons have a negative charge and are found in shells around the nucleus.
✓ The number of outer electrons in an atom depends on its group in the Periodic Table.

LEARNING OBJECTIVES

✓ Describe the formation of positive ions, known as cations, and negative ions, known as anions.
✓ State that an ionic bond is a strong electrostatic attraction between oppositely charged ions.
✓ Describe the formation of ionic bonds between elements from Group I and Group VII, including the use of dot-and-cross diagrams.
✓ Describe the properties of ionic compounds: high melting points and boiling points; good electrical conductivity when aqueous or molten and poor when solid (and generally soluble in water).
✓ **SUPPLEMENT** Describe the giant lattice structure of ionic compounds as a regular arrangement of alternating positive and negative ions, exemplified to sodium chloride.
✓ **SUPPLEMENT** Describe the formation of ionic bonds between ions of metallic and non-metallic elements, including the use of dot-and-cross diagrams.
✓ **SUPPLEMENT** Explain in terms of structure and bonding the properties of ionic compounds: high melting points and boiling points; good electrical conductivity when aqueous or molten and poor when solid.

THE FORMATION OF IONS

Atoms bond with other atoms in a **chemical reaction** to make a compound. For example, sodium reacts with chlorine to make sodium chloride. **Ionic compounds** contain a metal combined with one or more non-metals. They are not made up of molecules – they are made up of **ions**.

Ions are formed from atoms by the gain or loss of electrons. Both metals and non-metals try to achieve complete (filled) outer electron shells or the electron configuration of the nearest noble gas.

Metals lose electrons from their outer shells and form positive ions known as **cations**. Non-metals gain electrons in their outer shells and form negative ions known as **anions**.

ELECTRON TRANSFER IN IONIC BONDING

The **ionic bond** is a strong **electrostatic force** of attraction between these oppositely charged ions.

The bonding process can be represented in **dot-and-cross diagrams**. Dots show the electrons in one atom and crosses show the electrons in the other atom. Look at the reaction between sodium and chlorine as an example.

Sodium is a metal. It has a proton number of 11 and so has 11 electrons, arranged 2,8,1. Its atom diagram looks like this:	Chlorine is a non-metal. It has a proton number of 17 and so has 17 electrons, arranged 2,8,7. Its atom diagram looks like this:

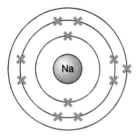 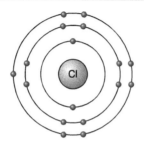

△ Fig. 2.13 Dot-and-cross diagrams for sodium and chlorine.

Sodium has one electron in its outer shell. It can achieve a full outer shell by losing this electron. The sodium atom transfers its outermost electron to the chlorine atom.	Chlorine has seven electrons in its outer shell. It can achieve a full outer shell by gaining an extra electron. The chlorine atom accepts an electron from the sodium.

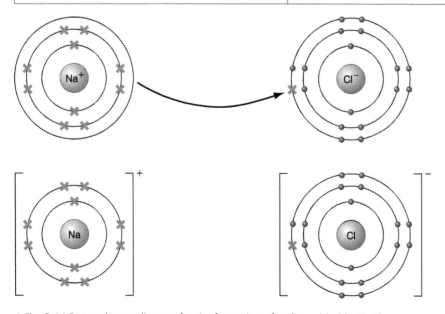

△ Fig. 2.14 Dot-and-cross diagram for the formation of sodium chloride, NaCl.

| The sodium is no longer an atom; it is now an ion. It does not have equal numbers of protons and electrons, so it is no longer neutral. It has one more proton than it has electrons, so it is a positive ion with a charge of 1+. The positive charge on the ion is shown outside the square brackets as +. The ion is written as Na^+. | The chlorine is no longer an atom; it is now an ion. It does not have equal numbers of protons and electrons, so it is no longer neutral. It has one more electron than it has protons, so it is a negative ion with a charge of 1−. The negative charge on the ion is shown outside the square brackets as −. The ion is written as Cl^-. |

Metals can transfer more than one electron to a non-metal

Magnesium combines with oxygen to form magnesium oxide. The magnesium (electron configuration 2,8,2) transfers two electrons to the oxygen (electron configuration 2,6). Magnesium therefore forms an Mg^{2+} ion and oxygen forms an O^{2-} ion.

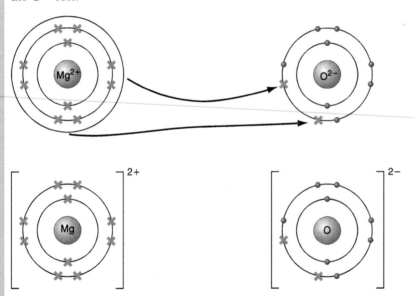

△ Fig. 2.15 Dot-and-cross diagram for the formation of magnesium oxide, MgO.

Aluminium has an electron configuration 2,8,3. When it combines with fluorine with an electron configuration 2,7, three fluorine atoms are needed for each aluminium atom. The formula of aluminium fluoride is therefore AlF_3.

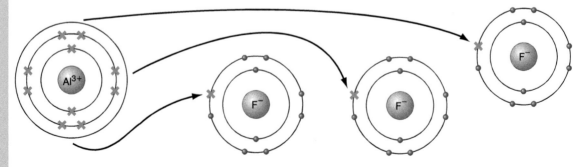

△ Fig. 2.16 Dot-and-cross diagram for aluminium fluoride, AlF_3.

1. What is a cation?

2. Draw a dot-and-cross diagram to show how lithium and fluorine atoms combine to form lithium fluoride. You must show the starting atoms and the finishing ions. (Proton numbers: Li 3; F 9)

3. SUPPLEMENT Draw a dot-and-cross diagram to show how calcium and sulfur atoms combine to form calcium sulfide. You must show the starting atoms and finishing ions. (Proton numbers: Ca 20; S 16)

4. SUPPLEMENT Consider the nature of the two elements phosphorus and oxygen and explain why you know that phosphorus oxide is not an ionic compound.

Electronic configuration and ionic charge

When atoms form ions, they are trying to achieve the electronic configuration of their nearest noble gas. Some common ions and their electronic configurations are shown in Table 2.6.

Ion	Electronic configuration
Li^+	2
Na^+	2,8
Mg^{2+}	2,8
F^-	2,8
Cl^-	2,8,8
O^{2-}	2,8

Δ Table 2.6 Electronic configurations of some ions.

PROPERTIES OF IONIC COMPOUNDS

Ionic compounds have a number of common properties:

a) high melting points and high boiling points

b) good electrical conductivity when in aqueous solution or when molten, and poor when solid.

c) Generally soluble in water (they dissolve to form solutions)

Ionic compounds form giant lattice structures of alternating positive and negative ions. It is called a giant structure of ions because there could be billions of ions held together in the lattice. A lattice is an ordered structure. For example, when sodium chloride is formed by ionic bonding, the ions do not pair up. Each sodium ion is surrounded by six chloride ions, and each chloride ion is surrounded by six sodium ions.

In these giant lattice structures the electrostatic attractions between the ions are very strong. This strong electrostatic attraction explains why ionic compounds have high melting and boiling points. For melting and boiling to occur the ions have to have sufficient energy to move apart and eventually to separate.

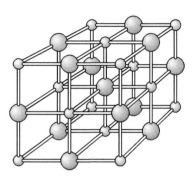

● chloride ion   ○ sodium ion

△ Fig. 2.18 In solid sodium chloride, the ions are held firmly in place. All ionic compounds have giant ionic lattice structures similar to this. This diagram shows only a small part of the repeating structure.

△ Fig. 2.17 Crystals of sodium chloride.

Properties of sodium chloride	Explanation in terms of structure
Hard crystals	Strong forces of attraction between the oppositely charged ions
High melting point (801 °C)	Strong forces of attraction between the oppositely charged ions
Dissolves in water	The water is also able to form strong electrostatic attractions with the ions – the ions are 'plucked' off the lattice structure
Does not conduct electricity when solid	Strong forces between the ions prevent them from moving
Conducts electricity when molten or dissolved in water	The strong forces between the ions have been broken down and so the ions are able to move

△ Table 2.7 Properties of sodium chloride.

Magnesium oxide is another ionic compound. It contains the ions Mg^{2+} and O^{2-}, and its ionic formula is MgO.

MgO has a much higher melting point and boiling point than NaCl because of the increased charges on the ions. The forces holding the ions together are stronger in MgO than in NaCl.

SCIENCE
IN
CONTEXT
MAGNESIUM OXIDE

Magnesium oxide is a compound with many uses. It is widely used in the construction industry, both in making cement and in making fire-proof construction materials. The fact that it has a melting point of over 2800 °C makes it ideal for this use.

Challenge Question: Like magnesium oxide, aluminium oxide has a melting point above 2000 °C. Suggest why aluminium oxide's melting point is so high.

△ Fig. 2.19 The heat resistance of magnesium oxide means that it is used to line furnaces.

SCIENCE
IN
CONTEXT
IONIC CRYSTALS

△ Fig. 2.20 Gemstones are examples of ionic crystals.

All ionic compounds form giant structures, and all have relatively high melting and boiling points. The charges on the ions determine the strength of the electrostatic attraction between the ions, and hence the melting and boiling points of the compound compared to others.

Another factor that affects the strength of the electrostatic attraction is the relative sizes of the positive and negative ions and how well they are able to pack together. The overall arrangement of the ions is determined by attractive forces between oppositely charged ions and repulsive forces between similarly charged ions. In sodium chloride, for example, six chloride ions fit around one sodium ion without the chloride ions getting too close together and repelling one another. Similarly, six sodium ions can fit around one chloride ion. This structure is sometimes called a 6:6 lattice (see Fig. 2.18 for a diagram of this structure).

Caesium is a metal in the same group of the Periodic Table as sodium, but caesium ions are much bigger than sodium ions. In the structure of caesium chloride, eight chloride ions can fit around each caesium ion. So although sodium and caesium are in the same group, their chlorides have different structures.

Some of the most valuable gemstones are ionic compounds. Rubies and sapphires, for example, are both aluminium oxide. The different colours of the gemstones are due to traces of other metals such as iron, titanium and chromium.

Challenge Question: Magnesium and calcium atoms both form ions with a charge of 2+. Suggest a reason why magnesium oxide and calcium oxide have different melting points.

QUESTIONS

1. Potassium chloride is a typical ionic compound. What are three properties you would expect potassium chloride to have?

2. SUPPLEMENT Explain why an ionic compound such as magnesium oxide does not conduct electricity when it is solid.

3. SUPPLEMENT Suggest a reason why magnesium oxide has a higher melting point than sodium chloride.

End of topic checklist

Key terms

anion, boiling point, cation, chemical reaction, dot-and-cross diagram, electrical conductivity, electrostatic attraction, ion, ionic bond, ionic compound, melting point, soluble

SUPPLEMENT giant lattice

During your study of this topic you should have learned:

◯ About the formation of ions by electron loss or gain.

◯ That an ionic bond is a strong electrostatic attraction between oppositely charge ions.

◯ How to describe the formation of ionic bonds between elements from Groups I and VII using dot-and-cross diagrams.

◯ That ionic compounds have high melting and boiling points and good electrical conductivity when aqueous or molten (and generally soluble in water).

◯ **SUPPLEMENT** How to describe the giant lattice structure of ionic compounds as a regular arrangement of alternating positive and negative ions.

◯ **SUPPLEMENT** How to describe the formation of ionic bonds between ions of metallic and non-metallic elements using dot-and-cross diagrams.

◯ **SUPPLEMENT** How to explain why ionic compounds have high melting and boiling points and good electrical conductivity when aqueous or molten.

End of topic questions

1. State which of the following reactions results in the formation of an ionic compound.

    A hydrogen and chlorine

    B carbon and hydrogen

    C sodium and oxygen

    D chlorine and oxygen

2. Using the Periodic Table on page 318, deduce the formulas of the ions formed by the following elements:

    a) potassium

    b) aluminium

    c) sulfur

    d) fluorine.

3. The table below shows the electronic arrangement of three atoms, X, Y and Z. Copy and complete the table to show the electronic arrangements and charges of the ions these atoms will form.

Atom	Electronic arrangement of the atom	Electronic arrangement of the ion	Charge on the ion
X	2,6		
Y	2,8,8,2		
Z	2,1		

4. Sketch a dot-and-cross diagram to show how potassium atoms (proton number 19) and fluorine atoms (proton number 9) combine to form potassium fluoride.

5. **SUPPLEMENT** Use a dot-and-cross diagram to show how the following atoms combine to form ionic compounds. (You must show the electronic arrangements of the starting atoms and the finishing ions.)

    a) potassium and oxygen (proton numbers: K 19; O 8)

    b) magnesium and chlorine (proton numbers: Mg 12; Cl 17)

6. **SUPPLEMENT** Explain why an ionic substance such as potassium chloride:

    a) has a high melting point

    b) can conduct electricity.

7. **SUPPLEMENT** Compare and contrast the structures of magnesium oxide and sodium chloride and explain why magnesium oxide has a higher melting point and boiling point than sodium chloride.

Molecules and covalent bonds

INTRODUCTION

Unlike ionic compounds, covalent compounds are formed when atoms of non-metals combine. Although covalent substances all contain the same type of bond, their properties can be quite different – some are gases, others are very hard solids with high melting points. Plastics are a common type of covalent substance. Because chemists now understand how the molecules form and link together, they can produce plastics with almost the perfect properties for a particular use, from soft and flexible (as in contact lenses) to hard and rigid (as in electrical sockets). The linkage of simple covalent molecules is covered in Chapter 11.

Δ Fig. 2.21 All plastics are covalent substances.

KNOWLEDGE CHECK

✓ Molecules are made up of two or more atoms combined together.
✓ The proton number is the number of protons in the nucleus of an atom.
✓ Electrons are found in shells around the nucleus of an atom.
✓ An element's group in the Periodic Table depends on the number of electrons in the outer shell of the atom.
✓ Noble gases have full outer electron shells.

LEARNING OBJECTIVES

✓ State that a covalent bond is formed when a pair of electrons is shared between two atoms leading to noble gas electronic configurations.
✓ Describe the formation of covalent bonds in simple molecules, including H_2, Cl_2, H_2O, CH_4, NH_3, and HCl. Use dot-and-cross diagrams to show the electronic configurations in these molecules.
✓ Describe in terms of structure and bonding the properties of simple molecular compounds: low melting points and boiling points; poor electrical conductivity.
✓ **SUPPLEMENT** Describe the formation of covalent bonds in simple molecules, including CH_3OH, C_2H_4, O_2, CO_2 and N_2. Use dot-and-cross diagrams to show the electronic configurations in these molecules.
✓ **SUPPLEMENT** Explain in terms of structure and bonding the properties of simple molecular compounds: low melting points and boiling points in terms of weak intermolecular forces (specific types of intermolecular forces are **not** required); poor electrical conductivity.
✓ Describe the giant covalent structures of graphite and diamond.
✓ **SUPPLEMENT** Relate the structures and bonding of graphite and diamond to their uses, limited to: graphite as a lubricant and as an electrode; diamond in cutting tools.

SIMPLE MOLECULES AND COVALENT BONDS

The non-metal atoms try to achieve complete outer electron shells or the electron configuration of the nearest noble gas by sharing electrons.

A single **covalent bond** is formed when two atoms each contribute one electron to a shared pair of electrons. For example, hydrogen gas exists as H_2 **molecules**. Each hydrogen atom needs to fill its electron shell. They can do this by sharing electrons. Both hydrogen atoms have now achieved noble gas electronic configurations.

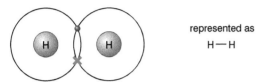

represented as

H—H

△ Fig. 2.22 The dot-and-cross diagram and displayed formula of H_2.

A covalent bond is the result of attraction between the bonding pair of electrons (negative charges) and the nuclei (positive charges) of the atoms involved in the bond. A single covalent bond can be represented by a single line. The formula of a hydrogen molecule can be written as a **displayed formula**, H—H.

The chlorine atoms in chlorine gas are also held together by single covalent bonds.

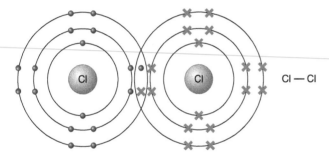

Cl — Cl

△ Fig 2.23 Dot-and-cross diagram and displayed formula for chlorine, Cl_2.

The hydrogen atoms and oxygen atoms in water are also held together by single covalent bonds.

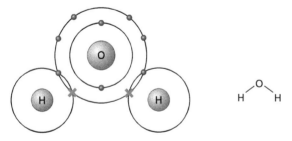

△ Fig. 2.24 Water contains single covalent bonds. All three atoms now have noble gas configurations.

The hydrogen atoms and carbon atoms in methane are held together by single covalent bonds.

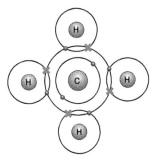

△ Fig. 2.25 Methane contains four single covalent bonds.

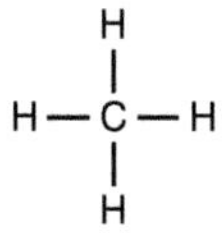

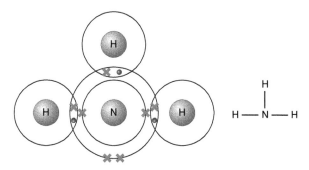

△ Fig. 2.26 The displayed formula for methane.

In ammonia the nitrogen atoms are joined to three hydrogen atoms by single covalent bonds.

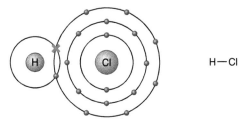

△ Fig 2.27 Dot-and-cross diagram and displayed formula for ammonia, NH_3.

The hydrogen chloride molecule, HCl, is also held together by a single covalent bond.

H—Cl

△ Fig. 2.28 Hydrogen chloride has a single covalent bond.

Ethane has a slightly more complex electron configuration.

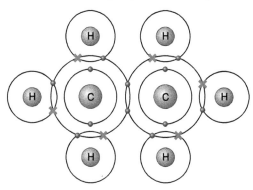

△ Fig. 2.29 Covalent bonds in ethane.

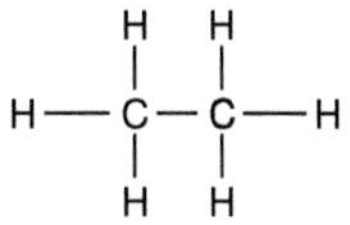

△ Fig. 2.30 Displayed formula for ethane.

The alcohol methanol is covalently bonded as shown in Fig. 2.31.

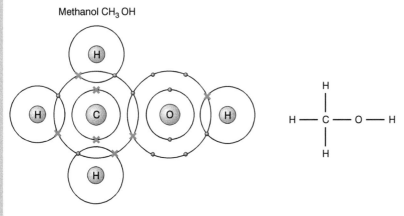

△ Fig. 2.31 Covalent bonds in methanol.

Each atom has achieved the configuration of a noble gas. Carbon and oxygen both now have 8 electrons in their outer shells and each hydrogen has two electrons in its outer shell. Some molecules contain double covalent bonds. In carbon dioxide, the carbon atom has an electron configuration of 2,4 and needs an additional four electrons to complete its outer electron shell. It needs to share its four electrons with four electrons from oxygen atoms (electron configuration 2,6). So two oxygen atoms are needed, each sharing two electrons with the carbon atom (so the formula is CO_2). In this way the carbon and oxygen atoms all have full outer electron shells, that is noble gas electronic configurations.

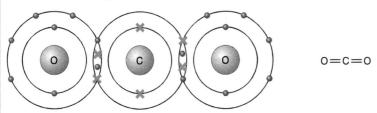

△ Fig. 2.32 Carbon dioxide contains double bonds.

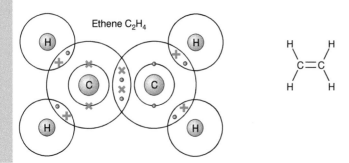

△ Fig. 2.33 Ethene contains a double bond.

In oxygen (O_2) there is also a double covalent bond as shown in the diagram below.

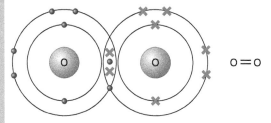

Δ Fig 2.34 Dot-and-cross diagram and displayed formula for oxygen O_2.

Some molecules contain triple covalent bonds. In the nitrogen molecule, each nitrogen atom has an electron configuration of 2,5 and needs an additional three electrons to complete its outer electron shell. It needs to share three of its outer electrons with another nitrogen atom. This forms a triple bond, which is shown as N ≡ N.

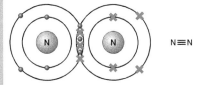

Δ Fig. 2.35 A nitrogen molecule contains a triple bond (each bond is formed from a pair of electrons).

QUESTIONS

1. Look at the dot-and-cross diagram for hydrogen (Fig 2.22). Why have both hydrogen atoms got noble gas electronic configurations?

2. Look at the dot-and-cross diagram for chlorine (Fig 2.23). How do you know from this diagram that the atomic number of chlorine is 17?

3. Look at the dot-and-cross diagram for ammonia (Fig 2.27) and explain why each nitrogen atom forms covalent bonds with three hydrogen atoms.

4. SUPPLEMENT Look at the dot-and-cross diagram for oxygen (Fig 2.34). Explain why the two oxygen atoms are linked together by a double covalent bond.

5. SUPPLEMENT Draw a dot-and-cross diagram and displayed formula to show the covalent bonds in hydrazine (N_2H_4). The proton number of hydrogen is 1; the proton number of nitrogen is 7.

How many covalent bonds can an element form?

The number of covalent bonds a non-metal atom can form is linked to its position in the Periodic Table. Metals (Groups I, II, III) do not form covalent bonds. The noble gases in Group VIII, for example helium, neon and argon, are unreactive and also do not usually form covalent bonds.

Group in the Periodic Table	I	II	III	IV	V	VI	VII	VIII (or 0)
Covalent bonds formed	–	–	–	4	3	2	1	–

△ Table 2.8 Group number and number of covalent bonds formed.

Properties of covalent compounds

Substances with molecular structures are usually gases, liquids or solids with low melting points and boiling points and poor electrical conductivity.

SUPPLEMENT

Covalent bonds are strong bonds. They are intramolecular bonds – formed *within* each molecule. Much weaker **intermolecular forces** attract the individual molecules to each other.

The properties of covalent compounds can be explained using a simple model involving these two types of bonds or forces.

Properties of hydrogen	Explanation in terms of structure
Hydrogen is a gas with a very low melting point (–259 °C).	The intermolecular forces of attraction between the molecules are weak.
Hydrogen does not conduct electricity.	There are no ions or free electrons present. The covalent bond (intramolecular bond) is a strong bond and the electrons cannot be removed from it easily.

△ Table 2.9 Properties of hydrogen.

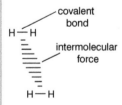

△ Fig. 2.36 Forces and bonds in and between hydrogen molecules.

Comparing the properties of covalent and ionic compounds

Simple covalent compounds typically have very different properties to ionic compounds. A comparison can be seen in Table 2.10. The volatility of a compound is a measure of how easily it forms a vapour.

Property	Ionic compounds	Simple covalent compounds
Solubility	Often soluble in water	Mostly insoluble in water
Electrical conductivity	Conduct electricity only when dissolved in water or molten (the ions separate and are free to move, carrying their electric charge)	Low electrical conductivity – are non-electrolytes (do not contain ions and so cannot carry an electrical current; however, some covalent compounds do form ions when dissolved in water)

Δ Table 2.10 Comparison of simple covalent compounds with ionic compounds.

SUPPLEMENT

Type of compound	Intermolecular force	Property
Ionic	Strong	High melting and boiling points
Simple covalent	Weak	Low melting and boiling points

QUESTIONS

1. SUPPLEMENT Explain why a covalently bonded compound such as carbon dioxide has a relatively low melting point.

2. SUPPLEMENT Suggest whether you expect a covalently bonded compound such as ethanol to conduct electricity. Explain your answer.

GIANT COVALENT STRUCTURES

Diamond and graphite

Some covalently bonded elements and compounds do not exist as simple molecular structures in the way that hydrogen does. Diamond, for example, has a **giant covalent structure** with each carbon atom covalently bonded to four others (Fig. 2.38). Another form of carbon is graphite. Graphite has a different giant structure, as seen in Fig. 2.39. Different forms of the same element, like these, are called allotropes. It is not necessary to learn this as it is not on the syllabus, but it provides useful context. In diamond, each carbon atom forms four strong covalent bonds. In graphite, each carbon atom forms three strong covalent bonds. There are weak forces of attraction between the layers in graphite (Fig. 2.39).

△ Fig. 2.37 A cut diamond is one of the hardest substances in nature.

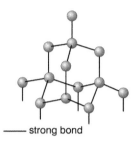

—— strong bond

△ Fig. 2.38 Structure of diamond.

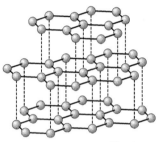

Structure of graphite

△ Fig. 2.39 Graphite is made of the same atoms as diamond but with a different molecular structure.

In diamond, all the bonding is extremely strong, which makes diamond an extremely hard substance – one of the hardest natural substances known. This is why diamonds are used in cutting tools.

In graphite, carbon atoms form layers of hexagons with strong covalent bonds. The weak forces of attraction are between the layers. Because the layers can slide over each other, graphite is flaky and can be used as a lubricant. Graphite can also conduct electricity, because the fourth unbonded electron from each carbon atom is **delocalised** and so can move along the layer. So graphite can be used as an electrode in electrical circuits.

The atoms in both diamond and graphite are held together by strong covalent bonds, which result in a very high melting point. Diamond has a melting point of about 3730 °C.

SCIENCE IN CONTEXT — DIAMONDS

Diamond is the hardest naturally occurring material in the world. It is formed by high pressures and high temperatures deep underground. Volcanic eruptions often bring the diamonds closer to the surface, where they can be mined. In some cases diamonds can be mined almost on the surface of the land, whereas in other cases tunnels need to be dug deep into the ground. Diamonds are mined throughout Africa.

The ore is called kimberlite, and diamonds are found in kimberlite gravels and pipe formations.

After mining, the ore is crushed, washed and screened by X-rays to find the diamonds. Finally the diamonds are sorted by hand, then washed and classified for sale. Diamond mining and recovery is a very clean operation. Processing of the ore uses no toxic chemicals and produces no chemical pollutants. However, getting the diamonds from deep in the ground can be very dangerous for the miners.

The quality of diamonds used in jewellery is judged in terms of the four Cs: carat weight, colour, clarity (how transparent the diamond is and how well it reflects light) and cut (the shape of the diamond). One carat is 0.2 g. Although the carat weight is the most important, prices can vary widely depending on the other three factors. In 2019 the largest producers were Russia, Australia and DR Congo, with Russia producing 19 million carats of uncut diamonds. It is estimated that Russia has reserves amounting to 650 million carats. Fig. 2.41 shows an open cast diamond mine in Yakutia, Russia. In the same year the world demand for diamonds was estimated at 39 billion dollars. Diamonds are clearly big business!

△ Fig. 2.40 This saw has diamonds on its cutting edges.

△ Fig. 2.41 An open cast diamond mine in Yakutia, Russia.

Challenge Question: Compare and contrast the properties of diamond and graphite and explain why the properties of diamond are so different from the properties of graphite.

QUESTIONS

1. Explain why diamond has a very high melting point.
2. Describe how the structure of graphite is different from that of diamond.

End of topic checklist

Key terms

covalent bond, delocalised electrons, displayed formula, giant covalent structure, molecule

SUPPLEMENT intermolecular forces

During your study of this topic you should have learned:

○ That a covalent bond is formed when a pair of electrons is shared between two atoms, leading to noble gas electronic configurations.

○ How to describe the formation of single covalent bonds in simple molecules, including H_2, Cl_2, H_2O, CH_4, NH_3 and HCl, using dot-and-cross diagrams.

○ How to describe in terms of structure and bonding the properties of simple molecular compounds, including low melting and boiling points and poor electrical conductivity.

○ How to describe the giant covalent structures of graphite and diamond.

○ **SUPPLEMENT** How to relate the structures and bonding of graphite and diamond to the use of graphite as a lubricant and electrode and the use of diamond in cutting tools.

○ **SUPPLEMENT** How to describe using dot-and-cross diagrams the formation of covalent bonds in simple molecules, including CH_3OH, C_2H_4, O_2, CO_2 and N_2.

○ **SUPPLEMENT** How to explain in terms of structure and bonding the properties of simple molecular compounds, including low melting and boiling points (weak intermolecular forces) and poor electrical conductivity.

End of topic questions

1. Identify which statement about diamond is **not** correct.

    A It has a very high melting point.

    B It is extremely hard.

    C It is a form of carbon.

    D It conducts electricity.

2. Draw dot-and-cross diagrams to show the bonding in the following compounds:

    a) hydrogen fluoride, HF

    b) SUPPLEMENT carbon disulfide, CS_2

    c) SUPPLEMENT ethanol, C_2H_5OH.

3. Ozone (O_3) is a gas found in the Earth's atmosphere. Explain how you know that ozone is covalently bonded and not ionically bonded.

4. SUPPLEMENT Candle wax is a covalently bonded compound. Explain why candle wax has a relatively low melting point.

5. SUPPLEMENT Explain why methane (CH_4), which has strong covalent bonds between the carbon atom and the hydrogen atoms, is a gas at room temperature and pressure and has a very low melting point.

6. SUPPLEMENT Substance X has a simple molecular structure.

    a) In which state(s) of matter would you predict X to exist in at room temperature and pressure? Explain your answer.

    b) Predict how you would expect the boiling point of X to compare with the boiling point of an ionic compound such as sodium chloride. Explain your answer.

7. SUPPLEMENT Use the structure of graphite to explain:

    a) how carbon fibres can add strength to tennis racquets

    b) how graphite conducts electricity.

8. **a)** SUPPLEMENT Diamond is probably the hardest naturally occurring material in the world. Explain this by referring to the structure of diamond.

    b) Apart from jewellery, name another use of diamond.

Metallic bonding

INTRODUCTION

The structure of a metal contains an orderly arrangement of atoms and can form crystals. The bonding in metals is different from that in ionic or covalent substances. As with ionic and covalent substances, the nature of the bonding in metals gives them their characteristic properties.

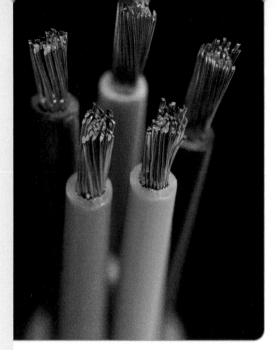

Δ Fig. 2.42 Copper atoms are held together by strong metallic bonds and form a giant structure.

KNOWLEDGE CHECK

✓ Protons, neutrons and electrons are important sub-atomic particles.
✓ Metals have characteristic properties.

LEARNING OBJECTIVES

✓ **SUPPLEMENT** Describe metallic bonding as the electrostatic attraction between the positive ions in a giant metallic lattice and a 'sea' of delocalised electrons.
✓ **SUPPLEMENT** Explain in terms of structure and bonding the properties of metals: good electrical conductivity and malleability.

The structure of metals

Metals are giant structures with high melting and boiling points. Metal atoms give up one or more of their electrons to form positive ions called **cations**. The electrons they give up form a 'sea' of electrons surrounding the positive metal ions. The negative electrons are attracted to the positive ions, holding the structure together.

The electrons are free to move through the whole structure. The electrons are **delocalised**, meaning they are not fixed in one position. This is **metallic bonding**.

The properties of metals

Metals are shiny, **malleable** (can be hammered into a sheet), have good electrical conductivity and good thermal conductors.

Metals are good conductors because their delocalised electrons are free to move through the structure. When a metal is in an electric circuit, electrons can move toward the positive terminal and the negative terminal can supply electrons to the metal.

Metals are malleable because metallic bonds are not as rigid as the bonds in diamond, for example, although they are still very strong. So the ions in the metal can move around into different positions when the metal is hammered or worked.

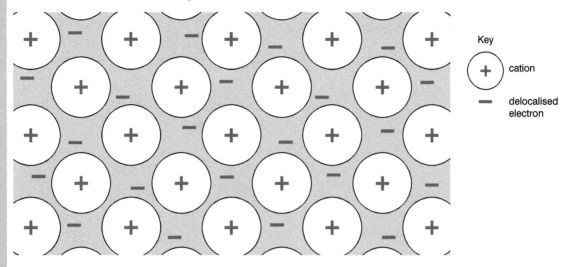

Key

+ cation

— delocalised electron

△ Fig. 2.43 Cations and delocalised electrons in a metal.

QUESTIONS

1. SUPPLEMENT What is a cation?

2. SUPPLEMENT Explain why metals are good conductors of electricity.

3. SUPPLEMENT Study the information above and describe what happens to the structure of a metal when you bend it.

SCIENCE IN CONTEXT   FACTS ABOUT METALS

1. The use of metals can be traced back to about 7000 years ago. An archaeological site in Serbia has evidence of the extraction of copper about that time, and gold artefacts dating to about 1000 years later have been found at a burial site at Varna in Bulgaria. In the period up to 2700 years ago, seven metals were known and used. These so-called 'metals of antiquity' were gold, copper, silver, lead, tin, iron and mercury. Of these gold, silver, copper, iron and mercury were found in their native state – that is, as pure elements. (Iron as a pure metal is found only in meteors.)

2. Mercury is the only liquid metal at normal room temperature and pressure.

3. The most reactive metals are found in Group I of the Periodic Table and include sodium and potassium.

4. Many of the metals used in construction are found in the middle of the Periodic Table and are called the transition metals.

5. Most of the metallic structures around us are not made from pure metals but from alloys. Alloys are mixtures of metals or occasionally mixtures of metals and non-metals: for example, steel is an alloy of iron and carbon. 39 000 tonnes of steel were used in building the Burj Khalifa building in Dubai (the world's tallest skyscraper at 828 metres). Alloys allow the properties of a metal to be modified for a particular purpose. For instance, aluminium is useful in building aircraft because it has a low density, but alloying it with other metals increases its strength. Bronze, an alloy of copper and tin, was the first alloy invented.

△ Fig. 2.44 The Burj Khalifa building in Dubai.

Challenge Question: Describe what properties of copper make it suitable for use in electrical wiring.

End of topic checklist

Key terms

SUPPLEMENT cation, conductor, delocalised electrons, malleability, metallic bonding

During your study of this topic you should have learned:

○ **SUPPLEMENT** That metallic bonding is the electrostatic attraction between the positive ions in a giant metallic lattice and a 'sea' of delocalised electrons.

○ **SUPPLEMENT** How to explain in terms of structure and bonding the following properties of metals: good electrical conductivity and malleability.

1. **SUPPLEMENT** Identify which statement about metals is **not** correct.

    A They all have high melting points.

    B They are good conductors of electricity.

    C They have a structure made up of positive ions in a giant metallic lattice.

    D They have a structure involving delocalised electrons.

2. **SUPPLEMENT** Use your knowledge of the structure of a metal to explain why metals:

    a) conduct electricity

    b) can be beaten into sheets (they are malleable).

3. **SUPPLEMENT** Analyse the information in the table. Use the information to answer the questions that follow.

Substance	Melting point (°C)	Boiling point (°C)	Electrical conductivity	
			Solid	Molten
A	751	1244	Poor	Good
B	−50	148	Poor	Poor
C	630	1330	Good	Good
D	247	696	Poor	Poor

Identify which substance:

a) is a metal

b) contains ionic bonds

c) has a giant covalent structure

d) has a simple molecular structure.

This section will give you plenty of practice at writing word and symbol equations. You will also see the importance of balancing equations so that the number and type of atoms in the substances that are reacting is equal to the number and type of atoms in the products formed so that matter is neither created nor destroyed in the reaction. You will then see how symbol equations enable you to work out the quantities of reactants needed to produce a certain amount of the products. These quantitative aspects of chemistry are crucially important in the chemical industry.

Starting points

1. Do you know the chemical formulas of common substances such as water and carbon dioxide? Make a list if you think you know more.

2. Can you remember a word equation you have come across before?

3. What do you think 'state symbols' might be?

4. Do you know what units can be used when writing down the concentration of a solution, for example, a solution of sodium chloride?

SYLLABUS SECTIONS COVERED

3.1 Formulas

3.2 Relative masses of atoms and molecules

3.3 The mole and the Avogadro constant

3

Stoichiometry

△ Industrial chemistry embraces technology.

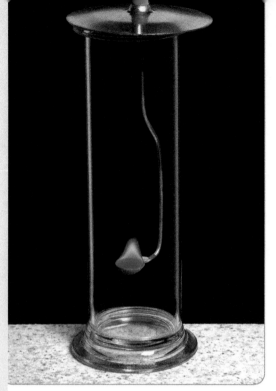

Formulas

Stoichiometry is the branch of chemistry concerned with the relative quantities of reactants and products in a chemical reaction. A study of stoichiometry depends on balanced chemical equations which, in turn, depend on knowledge of the chemical symbols for the elements and the formulas of chemical compounds. This topic starts by considering how simple chemical formulas are written and then looks in detail at chemical equations. The topic then focuses on how chemical equations can be used to work out how much reactant is needed to make a certain amount of product.

Δ Fig. 3.1 When this reaction is described as $S(s) + O_2(g) \rightarrow SO_2(g)$, it is understood by chemists all over the world.

KNOWLEDGE CHECK

✓ Elements are made up of atoms.
✓ Compounds are formed when atoms combine together.
✓ Molecules are formed in covalent bonding and ions are formed in ionic bonding.

LEARNING OBJECTIVES

✓ State the formulas of the elements and compounds named in the subject content.
✓ Define the molecular formula of a compound as the number and type of different atoms in one molecule.
✓ Deduce the formula of a simple compound from the relative number of atoms present in a model or a diagrammatic representation.
✓ Construct word equations to show how reactants form products.
✓ Balance and interpret simple symbol equations, including state symbols.
✓ SUPPLEMENT Deduce the formula of an ionic compound from the relative numbers of the ions present in a model or a diagrammatic representation or from the charges on the ions.
✓ SUPPLEMENT Construct symbol equations with state symbols, including ionic equations.
✓ SUPPLEMENT Deduce the symbol equation with state symbols for a chemical reaction, given relevant information.
✓ Describe relative atomic mass, A_r, as the average mass of the isotopes of an element compared to 1/12th of the mass of an atom of ^{12}C.
✓ Define relative molecular mass, M_r, as the sum of the relative atomic masses. Relative formula mass, M_r, will be used for ionic compounds.
✓ Calculate reacting masses in simple proportions Calculations will not involve the mole concept.
✓ State that concentration can be measured in g/dm^3.
✓ SUPPLEMENT State that the mole, mol, is the unit of amount of substance and that one mole contains 6.02×10^{23} particles, e.g. atoms, ions, molecules: this number is the Avogadro constant.

✓ **SUPPLEMENT** Use the relationship

$$\text{amount of substance (mol)} = \frac{\text{mass (g)}}{\text{molar mass (g/mol)}}$$

to calculate: amount of substance; mass; molar mass; relative atomic mass or relative molecular/formula mass.

✓ **SUPPLEMENT** Use the molar gas volume, taken as 24 dm³ at room temperature and pressure, r.t.p., in calculations involving gases.

✓ **SUPPLEMENT** Calculate stoichiometric reacting masses, limiting reactants, volumes of gases at r.t.p., including conversion between cm³ and dm³.

FORMULAS

When elements chemically combine, they form compounds. A compound can be represented by a **chemical formula**. The formula of a compound shows how many of each type of atom are present.

Where the compound contains only non-metals, and so is covalently bonded, the chemical formula is also known as a **molecular formula**.

All substances are made up from simple building blocks called elements. Each element has a unique **chemical symbol**, containing one or two letters. Elements discovered a long time ago often have symbols that don't seem to match their name. For example, silver has the chemical symbol Ag. This is derived from *argentum*, the Latin name for silver.

'Combining powers' of elements

There are a number of ways of working out chemical formulas. In this topic you will start with the idea of a 'combining power' for each element and then look at how the charges on ions can be used for ionic compounds. Later in the course you will be introduced to oxidation numbers and how these can be used to work out chemical formulas.

There is a simple relationship between an element's *group number* in the Periodic Table and its combining power. Groups are the vertical columns in the Periodic Table. The combining power is linked to the *number of electrons* in the outer shell of atoms of the element.

Group number	I	II	III	IV	V	VI	VII	VIII (or 0)
Combining power	1	2	3	4	3	2	1	0

△ Table 3.1 Combining powers of elements.

Groups I–IV: combining power = group number

Groups V–VII: combining power = 8 − (group number)

If an element is not in one of the main groups, its combining power is included in the name of the compound containing it. For example, copper is a transition metal and is in the middle block of the Periodic Table. In copper(II) oxide, copper has a combining power of 2.

Sometimes an element does not have the combining power you would predict from its position in the Periodic Table. The combining power of these elements is also included in the name of the compound containing it. For example, phosphorus is in Group V, so you would expect it to have a combining power of 3, but in phosphorus(V) oxide its combining power is 5.

The only exception is hydrogen. Hydrogen is not included in a group, nor is its combining power given in the name of compounds containing hydrogen. It has a combining power of 1.

Simple compounds

Many compounds contain just two elements. For example, when magnesium burns in oxygen, a white ash of magnesium oxide is formed. To work out the chemical formula of magnesium oxide:

1. Write down the name of the compound.
2. Write down the chemical symbols for the elements in the compound.
3. Use the Periodic Table to find the 'combining power' of each element. Write the combining power of each element under its symbol.
4. If the numbers can be cancelled down, do so.
5. Swap the combining powers. Write them after the symbol, slightly below the line (as a 'subscript').
6. If any of the numbers are 1, you do not need to write them.

Magnesium oxide has the chemical formula you would have probably guessed: MgO.

The chemical formula of a compound is not always immediately obvious, but if you follow these rules you will have no problems.

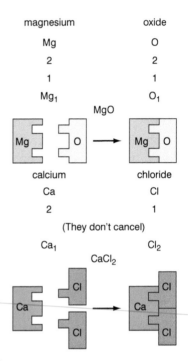

△ Fig. 3.2 Working out the chemical formulas for magnesium oxide and calcium chloride.

Compounds containing more than two elements

Some ionic compounds contain more than two elements. A small group of atoms bonded together can gain or lose electrons to give an overall charge to the group of atoms.

For example sodium sulfate, Na_2SO_4, contains sodium ions (Na^+) and sulfate ions (SO_4^{2-}).

Table 3.2 shows the names and formulas for some other ions that contain more than one element.

Name of ion	Formula of ion
Sulfate	SO_4^{2-}
Nitrate	NO_3^-
Carbonate	CO_3^{2-}
Hydroxide	OH^-

△ Table 3.2 Combining powers of common radicals.

The same rules for working out formulas apply to radicals as to elements. For example:

Copper(II) sulfate		Potassium nitrate	
Cu	SO_4	K	NO_3
2	2	1	1
$CuSO_4$		KNO_3	

△ Table 3.3 Combining elements and radicals.

If the formula contains more than one radical unit, the radical must be put in brackets. For example:

Calcium hydroxide	$Ca(OH)_2$
Ca	OH
2	1
$Ca(OH)_2$	

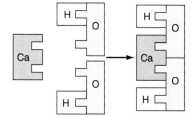

△ Fig. 3.3 Working out the chemical formula for calcium hydroxide.

The brackets are used just as they are used in mathematics: the number outside a bracket multiplies everything inside it. Be careful how you use the brackets – for example, do not be tempted to write calcium hydroxide as $CaOH_2$ rather than $Ca(OH)_2$. $CaOH_2$ is wrong.

$CaOH_2$ contains one Ca, one O, two H ✗
$Ca(OH)_2$ contains one Ca, two O, two H ✓

SUPPLEMENT

When you learned about ionic bonds you saw that atoms of some elements can transfer one or more outer shell electrons to atoms of other elements. Two ions are formed. The charge on the ion depends on how many electrons are gained or lost.

The formula of an ionic compound can be worked out from the ions present. For example, sodium chloride is an ionic compound.

Sodium is in Group I and forms an ion with a charge of 1+, Na^+. Chlorine is in Group VII and forms an ion with a charge of 1–, Cl^-. When these ions combine, the charges must cancel each other out: NaCl (the 1+ and 1– charges cancel).

What is the formula of lead(II) bromide, which contains Pb^{2+} and Br^- ions?

To cancel the 2+ charge, two 1– charges are needed, so the formula is $PbBr_2$.

QUESTIONS

1. How many atoms are there of each element in the following formulas: HCl; F_2; CH_3; H_2O; LiOH; $MgCl_2$; $CaCO_3$?

2. Deduce the chemical formulas of the following compounds:

    a) potassium bromide

    b) calcium oxide

    c) aluminium chloride

    d) carbon hydride (methane).

3. Deduce the chemical formulas of the following compounds:

    a) Deduce the formula of butane shown in the diagram

| Butane | $\underset{\displaystyle H-\overset{\displaystyle H}{\underset{\displaystyle H}{C}}-\overset{\displaystyle H}{\underset{\displaystyle H}{C}}-\overset{\displaystyle H}{\underset{\displaystyle H}{C}}-\overset{\displaystyle H}{\underset{\displaystyle H}{C}}-H}{}$ | Gas |

    b) Deduce the formula of propene shown in the diagram

| Propene | H—C—C=C (with H atoms) | Gas |

4. **SUPPLEMENT** Suggest the chemical formulas of the following compounds:

    a) a compound containing Zn^{2+} ions and Cl^- ions

    b) a compound containing Cr^{3+} ions and O^{2-} ions

    c) a compound containing Fe^{2+} ions and OH^- ions.

5. Suggest the formula of the ionic compound shown in the diagram.

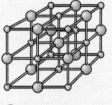

○ chloride ion ○ sodium ion

WRITING CHEMICAL EQUATIONS

In a chemical equation the starting chemicals are called the **reactants** and the finishing chemicals are called the **products**.

Follow these rules to write a chemical equation.

1. Write down the word equation.
2. Write down the symbols (for elements) and formulas (for compounds).
3. Balance the equation, to make sure there are the same number of each type of atom on each side of the equation.
4. Include the **state symbols**: solid (s); liquid (l); gas (g); solution in water (aq).

State	State symbol
Solid	(s)
Liquid	(l)
Gas	(g)
Solution	(aq)

△ Table 3.4 States and their symbols.

Remember that some elements are **diatomic**. They exist as molecules containing two atoms.

Element	Formula
Hydrogen	H_2
Oxygen	O_2
Nitrogen	N_2
Chlorine	Cl_2
Bromine	Br_2
Iodine	I_2

△ Table 3.5 Some diatomic elements.

WORKED EXAMPLES

1. When a lighted splint is put into a test-tube of hydrogen, the hydrogen burns with a 'pop'. In fact the hydrogen reacts with oxygen in the air (the reactants) to form water (the product).
   Write the chemical equation for this reaction.

   Word equation: hydrogen + oxygen $\rightarrow$ water

   Symbols and formulas: H_2 + O_2 $\rightarrow$ H_2O

   Balance the equation: $2H_2$ + O_2 $\rightarrow$ $2H_2O$

   For every two molecules of hydrogen that react, one molecule of oxygen is needed and two molecules of water are formed.

   Adding the state symbols: $2H_2(g) + O_2(g) \rightarrow 2H_2O(l)$

2. What is the equation when when sulfur reacts with oxygen in the air to form sulfur dioxide?

   Word equation: sulfur + oxygen $\rightarrow$ sulfur dioxide

   Symbols and formulas: S   + O_2 $\rightarrow$ SO_2

   Balance the equation: S   + O_2 $\rightarrow$ SO_2

   Adding the state symbols: S(s)   + $O_2(g)$ $\rightarrow$ $SO_2(g)$

Balancing equations

Balancing equations can be quite tricky. It is essentially done by trial and error. However, the golden rule is that *balancing numbers can only be put in front of the formulas.*

For example, to balance the equation for the reaction between methane and oxygen:

	Reactants	Products
Start with the unbalanced equation	$CH_4 + O_2$	$CO_2 + H_2O$
Count the number of atoms on each side of the equation	1C ✓, 4H, 2O	1C ✓, 2H, 3O
There is a need to increase the number of H atoms on the products side of the equation. Put a '2' in front of the H_2O	$CH_4 + O_2$	$CO_2 + 2H_2O$

	Reactants	Products
Count the number of atoms on each side of the equation again	1C ✓, 4H ✓, 2O	1C ✓, 4H ✓, 4O
There is a need to increase the number of O atoms on the reactant side of the equation. Put a '2' in front of the O_2	$CH_4 + 2O_2$	$CO_2 + 2H_2O$
Count the number of atoms on each side of the equation again	1C ✓, 4H ✓, 4O ✓	1C ✓, 4H ✓, 4O ✓

△ Table 3.6 Steps in balancing the equation for the reaction between methane and oxygen.

No atoms have been created or destroyed in the reaction. The equation is balanced.

$$CH_4(g) + 2O_2(g) \rightarrow CO_2(g) + 2H_2O(l)$$

△ Fig. 3.4 The number of each type of atom is the same on the left and right sides of the equation.

QUESTIONS

1. Balance the following chemical equations:

   a) $Ca(s) + O_2(g) \rightarrow CaO(s)$

   b) $H_2S(g) + O_2(g) \rightarrow SO_2(g) + H_2O(l)$

2. SUPPLEMENT Suggest balanced equations for the following word equations:

   a) carbon + oxygen → carbon dioxide

   b) magnesium + oxygen → magnesium oxide

   c) copper(II) oxide + hydrogen → copper + water

As mentioned earlier, the general method for balancing equations is by trial and error, but it helps if you are systematic – always start on the left-hand side with the reactants. Sometimes you can balance an equation using fractions. In more advanced study such balanced equations are perfectly acceptable. Getting rid of the fractions is not difficult though. Look at this example:

WORKED EXAMPLE

Ethane (C_2H_6) is a hydrocarbon fuel and burns in air to form carbon dioxide and water.

Unbalanced equation: $C_2H_6(g) + O_2(g) \rightarrow CO_2(g) + H_2O(l)$

Balancing the carbon and hydrogen atoms gives:

$$C_2H_6(g) \quad + \quad O_2(g) \quad \rightarrow \quad 2CO_2(g) \quad + \quad 3H_2O(l)$$

The equation can then be balanced by putting 3½ in front of the O_2. By doubling every balancing number, the equation is then balanced using whole numbers.

$$2C_2H_6(g) \quad + \quad 7O_2(g) \quad \rightarrow \quad 4CO_2(g) \quad + \quad 6H_2O(l)$$

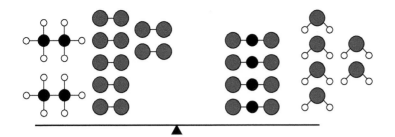

△ Fig. 3.5 Balancing the equation for burning ethane in air.

During your course you will become familiar with balancing equations and become much quicker at doing it. Try balancing the equations below. The third is the chemical reaction often used for making chlorine gas in the laboratory.

QUESTIONS

1. Balance the following equations:

   a) $C_5H_{10}(g) + O_2(g) \rightarrow CO_2(g) + H_2O(l)$

   b) $Fe_2O_3(s) + CO(g) \rightarrow Fe(s) + CO_2(g)$

   c) $KMnO_4(s) + HCl(aq) \rightarrow KCl(aq) + MnCl_2(aq) + H_2O(l) + Cl_2(g)$

SUPPLEMENT

Ionic equations

Ionic equations show reactions involving ions (atoms or groups of atoms that have lost or gained electrons). The size of the charge on an ion is the same as its combining power – whether it is positive or negative depends on which part of the Periodic Table the element is placed in.

In many ionic reactions some of the ions play no part in the reaction. These ions are called spectator ions. It is not necessary to learn this as it is not on the syllabus, but it provides useful context. A simplified ionic equation can be written, using only the important, reacting ions. In these equations, state symbols are often used and appear in brackets.

The equation must balance in terms of chemical symbols and charges.

WORKED EXAMPLES

1. When lead(II) nitrate solution is mixed with potassium iodide solution, the products are insoluble lead(II) iodide and potassium nitrate. The lead iodide forms a coloured **precipitate** (Fig. 3.6). The word equation is:

   lead(II) nitrate + potassium iodide → lead(II) iodide + potassium nitrate

   In the reaction, the potassium ions and nitrate ions remain separate – they are spectators. The important ions are the lead(II) ions Pb^{+2} and the iodide ions I^-.

	Reactants	Products
	$Pb^{2+}(aq) + 2I^-(aq)$	$PbI_2(s)$
Symbols	1 Pb ✓, 2I ✓	1 Pb ✓, 2I ✓
Charges	2+ and 2− = 0	0 ✓

△ Fig 3.6

The balanced ionic equation is:

$$Pb^{2+} (aq) + 2I^- (aq) \rightarrow PbI_2 (s)$$

The equation shows that *any* solution containing lead(II) ions will react with *any* solution containing iodide ions to form lead(II) iodide.

2. Any solution containing copper(II) ions (Cu^{2+}) and any solution containing hydroxide ions (OH^-) can be used to make copper(II) hydroxide ($Cu(OH)_2$), which appears as a solid precipitate (Fig. 3.7):

	Reactants	Products
	$Cu^{2+}(aq) + 2OH^-(aq)$	$Cu(OH)_2(s)$
Symbols	1Cu ✓, 2O ✓, 2H ✓	1Cu ✓, 2O ✓, 2H ✓
Charges	2+ and 2− = 0 ✓	0 ✓

The balanced ionic equation is:

$$Cu^{2+}(aq) + 2OH^-(aq) \rightarrow Cu(OH)_2(s)$$

△ Fig. 3.7 Copper(II) hydroxide.

Reacting quantities

Observing chemical reactions can be fun. Some can be quite dramatic, producing sparks or flames; others might produce unexpected colour changes. However, many of the reactions you might see in the laboratory are used on an industrial scale to make chemicals for manufacturing other products or in everyday living. So it is not enough to know *what* can be made in a reaction; scientists also need to know *how much* can be made from the quantities of starting materials available. To do this we need to be able to weigh atoms, or do we?

RELATIVE MASSES OF ATOMS AND MOLECULES

Relative atomic mass

Atoms are far too light to be weighed. Instead, scientists have developed a **relative atomic mass** scale. The lightest atom, hydrogen, was chosen at first as the unit that all other atoms were weighed against.

△ Fig. 3.8 What volume of chlorine was needed to make this bottle of bleach?

On this scale, a carbon atom weighs the same as 12 hydrogen atoms, so carbon's relative atomic mass was given as 12.

Using this relative mass scale you can see, for example, that:

- 1 atom of magnesium is 24 × the mass of 1 atom of hydrogen.
- 1 atom of magnesium is 2 × the mass of 1 atom of carbon.
- 1 atom of copper is 2 × the mass of 1 atom of sulfur.

	Hydrogen	Carbon	Oxygen	Magnesium	Sulfur	Calcium	Copper
Symbol	H	C	O	Mg	S	Ca	Cu
Relative atomic mass	1	12	16	24	32	40	64
Relative size of atom							

△ Table 3.7 Relative atomic masses and sizes of atoms.

Since 1961 the reference point of the relative atomic scale has been carbon-12 (^{12}C).

The relative atomic mass, A_r, is the average mass of naturally occurring atoms of an element on a scale where the ^{12}C atom has a mass of exactly 12 units. This takes into account the abundance of all existing isotopes of that element.

Relative formula masses, M_r

A **relative formula mass** (M_r) can be worked out from the **relative atomic masses** of the atoms in the formula.

The relative formula mass of a molecule (also known as the **relative molecular mass**) can be worked out by simply adding up the relative atomic masses of the atoms in the molecule. For example:

Water, H_2O (H = 1, O = 16)

The relative formula mass (M_r) = 1 + 1 + 16 = 18

Note: The subscript $_2$ only applies to the hydrogen atom.

Carbon dioxide, CO_2 (C = 12, O = 16)

M_r = 12 + 16 + 16 = 44

Note: The subscript $_2$ only applies to the oxygen atom.

A similar approach can be used for any formula, including ionic formulas. As ionic compounds do not contain molecules the term relative molecular mass cannot be used. That is why the term relative formula mass is used, as it can be applied to ionic and covalent compounds.

WORKED EXAMPLES

1. Sodium chloride, NaCl (A_r: Na = 23, Cl = 35.5)

   The relative formula mass (M_r) = 23 + 35.5 = 58.5

2. Potassium nitrate, KNO_3 (A_r: K = 39, N = 14, O = 16)

   M_r = 39 + 14 + 16 + 16 + 16 = 101

   (Note: The subscript $_3$ only applies to the oxygen atoms.)

3. Calcium hydroxide, $Ca(OH)_2$ (A_r: Ca = 40, O = 16, H = 1)

$M_r = 40 + (16 + 1) \times 2 = 40 + 34 = 74$

(Note: The subscript $_2$ applies to everything inside the bracket.)

4. Magnesium nitrate, $Mg(NO_3)_2$ (A_r: Mg = 24, N = 14, O = 16)

$M_r = 24 + (14 + 16 + 16 + 16) \times 2 = 24 + (62) \times 2 = 24 + 124 = 148$

QUESTIONS

1. Calculate the relative formula mass of methane, CH_4. (A_r: H = 1, C = 12)

2. Calculate the relative formula mass of ethanol, C_2H_5OH. (A_r: H = 1, C = 12, O = 16)

3. Calculate the relative formula mass of ozone, O_3. (A_r: O = 16)

Experiments to find the formulas of simple compounds

Magnesium ribbon can be heated in a crucible to make a white powder called magnesium oxide. The magnesium reacts with oxygen from the air. The reaction is called oxidation because the magnesium combines with oxygen.

If you measure the masses of the magnesium used and the magnesium oxide formed in this reaction, you can use the relative atomic masses of magnesium and oxygen to work out the formula of the compound made.

Measurement	Mass in grams (g)
Mass of crucible + lid	30.00
Mass of crucible + lid + magnesium	30.24
Mass of crucible + lid + magnesium oxide	30.40
Mass of magnesium	30.24 − 30.00 = 0.24
Mass of magnesium oxide	30.40 − 30.00 = 0.40
Mass of oxygen	30.40 − 30.24 = 0.16

△ Table 3.8 Typical results when 0.24 g of magnesium is oxidised.

The result is that 0.24 g of magnesium joins with 0.16 g of oxygen. Therefore 24 g of magnesium would join with 16 g of oxygen. The relative atomic mass of magnesium is 24 and that of oxygen is 16. So 1 magnesium atom combines with 1 oxygen atom. This means that the formula of magnesium oxide is MgO.

△ Fig. 3.9 Magnesium ribbon is put in a crucible with a lid on it. The crucible is heated until the magnesium is red hot. The lid is lifted very slightly (to allow oxygen in) and put back down. This lets the magnesium burn but prevents loss of magnesium oxide.

WORKED EXAMPLE

Copper(II) oxide can be heated in hydrogen to produce copper and water. This reaction can be used to find the formula of copper(II) oxide.

In an experiment 12.8 g of copper was produced from 16.0 g of copper(II) oxide.

Relative atomic masses: H = 1, O = 16, Cu = 64

Word equation: copper(II) oxide + hydrogen → copper + water

Masses   16.0 g   12.8 g

So the mass of oxygen that was combined with the copper
= 16.0 − 12.8 = 3.2 g.

Therefore 32 g of oxygen would combine with 128 g of copper.

Therefore $\dfrac{32}{16} = 2$

$\dfrac{128}{64} = 2$

Therefore the formula of copper(II) oxide is CuO.

Developing practical skills

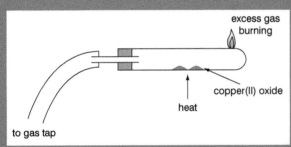

△ Fig. 3.10 Apparatus to find the chemical formula of a sample of copper(II) oxide. Safety note: Eye protection should be worn and care taken to ensure there are no gas leaks around the stopper and only a gentle flow of gas through the tube.

Two students set out to find the chemical formula of a sample of copper(II) oxide. First they weighed an empty combustion tube, then used a spatula to put a full measure of copper(II) oxide near the centre of the tube, being careful not to spill any of the powder near the opening of the tube. They then reweighed the tube and set up the apparatus as shown in Fig. 3.10. (Note: copper(II) oxide is harmful if swallowed.)

The gas was turned on very slowly and after about 10 seconds a lighted splint was held over the jet. The gas flow was adjusted until a flame about 1 cm high was burning at the jet. The copper(II) oxide was then heated strongly in a Bunsen flame for about 15 minutes until all the copper(II) oxide had been turned into copper. At this point the tube was allowed to cool, but with the gas still flowing through the tube and being burned at the jet.

When the tube was cold, the gas was turned off and the tube and contents reweighed. The results are shown in the table:

Mass of tube + copper(II) oxide	= 24.15 g
Mass of tube + copper	= 23.92 g
Mass of tube	= 23.15 g
Mass of copper(II) oxide	=
Mass of copper	=
Mass of oxygen combined with the copper	=

Using and organising techniques, apparatus and materials

1. Explain why it was important not to spill any of the copper(II) oxide near the opening of the tube.

2. Explain why the gas supply was turned on very slowly.

3. Suggest why the tube was allowed to cool before turning the gas supply off.

Observing, measuring and recording

4. Describe the colour change you would expect to see as the copper(II) oxide changes into copper.

5. Use the results to calculate the masses of copper(II) oxide, copper and oxygen.

Handling experimental observations and data

6. Use your results to work out the proportions of the relative atomic mass of copper and the relative molecular mass of oxygen.

7. Calculate the ratio of copper : oxygen. (Write this as Cu : O 1 : ?)

Evaluating methods

8. What do your calculations suggest is the most likely formula of copper(II) oxide?

9. Evaluate the procedure used and identify the main sources of error in the experiment.

QUESTIONS

1. In the experiment used to find the formula of magnesium oxide:

   a) Explain why the lid of the crucible is lifted whilst the magnesium is being heated.

   b) Explain why the crucible lid is lifted only very slightly during the heating of the magnesium.

   c) Suggest the colour of magnesium oxide.

THE MOLE AND THE AVOGADRO CONSTANT

The **mole**, mol, is the unit of amount of substance. It contains 6.02×10^{23} particles, for example, atoms, ions, or molecules. It is also known as the **Avogadro Constant** after the scientist who discovered it.

REMEMBER

When calculating moles of elements, you must be careful to make sure you know what the question refers to. For example, you may be asked for the mass of 1 mole of nitrogen gas. N = 14, but nitrogen gas is diatomic, that is N_2, so the mass of 1 mole of N_2 = 28 g. This applies to other diatomic elements, such as Cl_2, Br_2, I_2, O_2 and H_2.

QUESTIONS

1. **SUPPLEMENT** In an experiment, 5.6 g of iron reacts to form 8.0 g of iron oxide. Calculate the formula of the iron oxide. (A_r: O = 16; Fe = 56)

2. **SUPPLEMENT** 16.2 g of zinc oxide is reduced by carbon to form 13.0 g of zinc. Calculate the formula of zinc oxide. (A_r: O = 16; Zn = 65)

Moles

The mole concept can be difficult to understand. So it is worth just revising some key points.

A mole is an amount of substance. As already stated it is a very large number, approximately 6.02×10^{23}. This number is called the Avogadro number or Avogadro constant.

You can have a mole of atoms, a mole of molecules or a mole of electrons.

The mole is the amount of substance that contains 6.02×10^{23} particles. The particles can be atoms, molecules or ions.

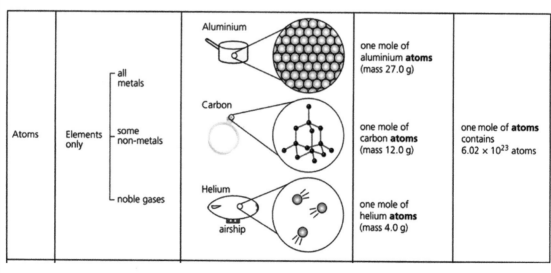

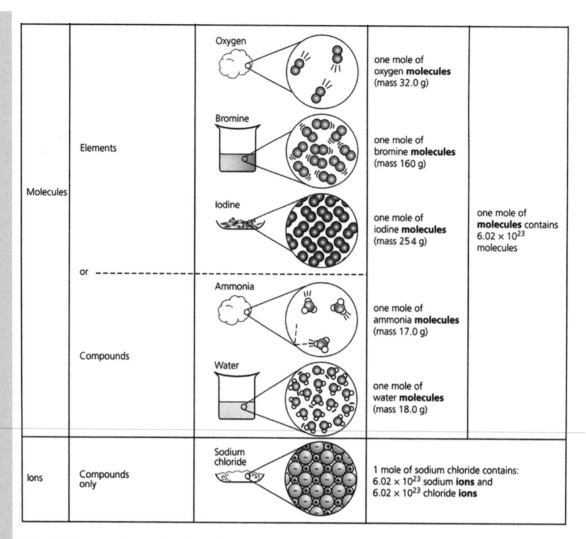

Molecules	Elements	Oxygen — one mole of oxygen **molecules** (mass 32.0 g) Bromine — one mole of bromine **molecules** (mass 160 g) Iodine — one mole of iodine **molecules** (mass 254 g)	one mole of **molecules** contains 6.02×10^{23} molecules
	or		
	Compounds	Ammonia — one mole of ammonia **molecules** (mass 17.0 g) Water — one mole of water **molecules** (mass 18.0 g)	
Ions	Compounds only	Sodium chloride — 1 mole of sodium chloride contains: 6.02×10^{23} sodium **ions** and 6.02×10^{23} chloride **ions**	

Δ Fig. 3.11 Moles of atoms, molecules and ions.

The relative atomic mass of an element tells you the mass of 1 mole of atoms of that element. So, for example, 1 mole of carbon atoms has a mass of 12 grams. The molar mass of carbon is 12 g/mol.

The relative formula mass (M_r) tells you the mass of 1 mole of that substance. For example, 1 mole of sodium chloride, NaCl, has a relative formula mass of 58.5, so 1 mole has a mass of 58.5 g.

WORKED EXAMPLES

1. How many moles of atoms are there in 72 g of magnesium?
   (A_r of magnesium = 24)

   Write down the formula:
   $$\text{amount of substance (mol)} = \frac{\text{mass (g)}}{\text{molar mass (g/mol)}}$$

   Rearrange if necessary: (None needed)

   Substitute the numbers:
   $$\text{moles} = \frac{72}{24}$$

   Write the answer and units: moles = 3 moles

2. What is the mass of 0.1 mole of carbon atoms? (A_r of carbon = 12)

Write down the formula:
$$\text{moles} = \frac{\text{mass}}{A_r}$$

Rearrange if necessary: mass = moles × molar mass

Substitute the numbers: mass = 0.1 × 12

Write the answer and units: mass = 1.2 g

Linking reactants and products

Chemical equations allow quantities of reactants and products to be linked together. They tell you how much of the products you can expect to make from a fixed amount of reactants.

In a balanced equation, the numbers in front of each symbol or formula indicate the numbers of moles represented. The numbers of moles can then be converted into masses in grams.

For example, when magnesium (A_r = 24) reacts with oxygen (A_r = 16):

Write down the balanced equation:
$2Mg(s) + O_2(g) \rightarrow 2MgO(s)$

Write down the number of moles: 2 + 1 → 2

Convert moles to masses: 48 g + 32 g → 80 g

So when 48 g of magnesium reacts with 32 g of oxygen, 80 g of magnesium oxide is produced. From this you should be able to work out the mass of magnesium oxide produced from any mass of magnesium.

△ Fig. 3.12 In an oxidation reaction, magnesium reacts with oxygen – the reaction can be used in fireworks and flares producing a brilliant white colour.

WORKED EXAMPLES

1. What mass of magnesium oxide can be made from 6 g of magnesium? (A_r: O = 16, Mg = 24)

Equation:	2Mg(s)	+	O₂(g)	→	2MgO(s)
Moles:	2		1		2
Masses:	48 g		32 g		80 g
÷ 8	6 g				10 g

Therefore 6 g of Mg will form 10 g of MgO.

Note: In this example, there was no need to work out the mass of oxygen needed. It was assumed that there would be as much as was necessary to convert all the magnesium to magnesium oxide.

2. What mass of ammonia can be made from 56 g of nitrogen? (A_r: H = 1, N = 14)

Equation:	N₂(g)	+	3H₂(g)	→	2NH₃(g)
Moles:	1		3		2
Masses:	28 g		6g		34 g
× 2	56 g				68 g

Mass of ammonia = 68 g

QUESTIONS

1. Calculate what mass of calcium oxide can be made from the decomposition of 50 g of calcium carbonate.
   (A_r: C = 12; O = 16; Ca = 40)

   Equation: $CaCO_3(s) \rightarrow CaO(s) + CO_2(g)$

2. When magnesium burns in oxygen, magnesium oxide is formed. Calculate what mass of magnesium oxide can be made from 6 g of magnesium, assuming there is as much oxygen as is needed. (A_r: Mg = 24, O = 16)

SUPPLEMENT

Moles of gases

In reactions involving gases it is often more convenient to measure the *volume* of a gas rather than its mass.

There are many gases and they are crucially important in science. In experiments or industrial processes it is often necessary to know the amount of a gas – but a gas is difficult to weigh. **Molar gas volumes** make it possible to find out the amount of a gas by using volume rather than mass.

One mole of any gas occupies the same volume under the same conditions of temperature and pressure. The conditions chosen are usually room temperature (25 °C) and normal atmospheric pressure (1 atmosphere).

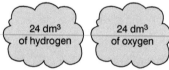

△ Fig. 3.13 Each of these contains 1 mole of molecules.

The volume of one mole of any gas contains the Avogadro number of molecules (particles) of that gas. This means that equal volumes of all gases at the same temperature and pressure must contain the same number of molecules. This is sometimes called **Avogadro's constant**.

One mole of any gas occupies 24 000 cm³ (24 dm³) at room temperature and pressure (r.t.p.). The following equation can be used to convert gas volumes into moles and vice versa:

$$\text{moles} = \frac{\text{volume in cm}^3}{24\,000}$$

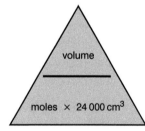

△ Fig. 3.14 The triangle can be used to decide whether to multiply or divide the quantities.

It is important to be able to convert dm³ into cm³ and vice versa. Remember that 1 dm³ is equivalent to 1000 cm³. So for example 0.5 dm³ = 500 cm³ and 1200 cm³ = 1.2 dm³.

QUESTIONS

1. SUPPLEMENT Calculate how many moles of molecules there are in the following:

   a) 88 g of carbon dioxide, CO_2. (A_r: C = 12; O = 16)

   b) 9 g of water, H_2O. (A_r: H = 1; O = 16)

   c) 2.8 g of ethene, C_2H_4. (A_r: H = 1; C = 12)

2. **SUPPLEMENT** Calculate how many moles of molecules there are in the following:

   a) 12 000 cm³ of sulfur dioxide, SO_2, at room temperature and pressure.

   b) 2400 cm³ of methane, CH_4, at room temperature and pressure.

   c) 48 000 cm³ of oxygen, O_2, at room temperature and pressure.

WORKED EXAMPLES

1. What volume of hydrogen is formed at room temperature and pressure when 4 g of magnesium is added to excess dilute hydrochloric acid? (A_r: H = 1, Mg = 24, molar volume at room temperature and pressure = 24 000 cm³)

   Balanced equation: $Mg(s) + 2HCl(aq) \rightarrow MgCl_2(aq) + H_2(g)$

   Moles: 1   2   1   1

   Masses/volumes: 24 g   24 000 cm³

   Number of moles of magnesium $= \dfrac{4}{24} = 0.167$

   Therefore 0.167 moles hydrogen gas will be produced.

   Therefore volume of hydrogen gas $= 0.167 \times 24\ 000$

   $$= 4000 \text{ cm}^3$$

   Note: The hydrochloric acid is in excess. This means that there is enough to react with all the magnesium.

2. What volume of carbon dioxide will be produced when 124 g of copper carbonate is broken down by heating?

   (A_r: C = 12, O = 16, Cu = 64, molar volume at room temperature and pressure = 24 000 cm³)

   Words: copper carbonate $\rightarrow$ copper oxide + carbon dioxide

   Balanced equation: $CuCO_3(s)$ $\rightarrow$ $CuO(s)$ + $CO_2(g)$

   Moles: 1   1   1

   Masses/vols: 124 g   24 000 cm³

   The relative formula mass of $CuCO_3$ is 64 + 12 + 16 + 16 + 16 = 124 g

   As there is exactly one mole of reactant, there must be one mole of each of the products. So, one mole of CO_2 is produced or 24 000 cm³.

AMEDEO AVOGADRO

Amedeo Avogadro was born in Italy in 1776. At age 20 he graduated from university with a degree in religious law. However, his real interests were science and mathematics, and this was where he started to devote his university research. At the beginning of the 19th century, scientists across the world were trying to work out the relative masses of different atoms. There was much confusion and disagreement about the existence of other particles, which we now call molecules. Against this background of confusion, in 1811 Avogadro published a research paper on 'determining the relative masses of elementary molecules'. It was in this paper that he proposed his hypothesis:

'All gases under the same conditions of temperature and pressure contain the same number of molecules.'

The scientific community paid little attention to his work at the time. There were doubts about the existence of molecules and conflicting evidence coming from the research of more well-known and respected scientists. In fact, it was not until four years after his death in 1856 that his idea was accepted, following work done by a highly respected scientist called Cannizzaro.

Δ Fig. 3.15 Amedeo Avogadro.

So Avogadro died without ever knowing the importance of his contribution to the development of science. It was only in 1911 that Avogadro's contribution to chemistry was formally recognised when, in honour of his work, the number of molecules in one mole of a substance was called the Avogadro number, equal to approximately 6×10^{23}.

Challenge Question: Consider other great advances or discoveries in science. Can you think of any that have not been recognised during the lifetime of the scientist concerned?

CONCENTRATION OF SOLUTIONS

A solution is made when a **solute** dissolves in a **solvent**. The **concentration** of a solution depends on how much solute is dissolved in how much solvent. The units of concentration are using given as g/dm^3.

End of topic checklist

Key terms

chemical formula, chemical symbol, concentration, diatomic, ionic equation, molar gas volumes, mole, molecular formula, precipitate, products, reactant, relative atomic mass (A_r), relative formula mass (M_r), relative molecular mass (M_r), spectator ions, state symbols

SUPPLEMENT Avogadro constant, molar gas volume, molar mass, mole

During your study of this topic you should have learned:

◯ How to use the symbols of the elements to write the formulas of simple compounds.

◯ How to define the molecular formula of a compound.

◯ How to deduce the formula of a simple compound from the numbers of atoms present.

◯ How to deduce the formula of a simple compound from a model or a diagram.

◯ How to construct word equations and simple balanced symbol equations, including state symbols.

◯ That the relative atomic mass (A_r) is the average mass of the isotopes of an element compared to 1/12th of the mass of an atom of ^{12}C.

◯ That relative molecular mass (M_r) is the sum of the relative atomic masses.

◯ That relative formula mass (M_r) is used for ionic compounds.

◯ How to use simple proportion to work out reacting masses.

◯ That solution concentrations can be measured in g/dm^3 or mol/dm^3.

◯ How to recall the formulas of some common elements and compounds, including hydrogen, water, carbon dioxide, ammonia, methane, sodium chloride and hydrochloric acid.

◯ **SUPPLEMENT** How to determine the formula of an ionic compound from the relative number of ions present in a model or diagram or from the charges on the ions present.

◯ **SUPPLEMENT** How to construct equations with state symbols, including ionic equations.

◯ **SUPPLEMENT** How to deduce the balanced equation for a chemical reaction, given relevant information.

○ **SUPPLEMENT** That the mole (mol) is the unit of the amount of substance and that one mole contains 6.02×10^{23} particles (atoms, molecules, ions). This is the Avogadro constant.

○ **SUPPLEMENT** How to use the relationship

$$\text{amount of substance (mol)} = \frac{\text{mass (g)}}{\text{molar mass (g / mol)}}$$

○ **SUPPLEMENT** How to use the molar gas volume (24 dm^3 at room temperature and pressure, r.t.p.).

○ **SUPPLEMENT** How to calculate stoichiometric reacting masses, limiting reactants, volumes of gases at r.t.p., volumes of solutions.

End of topic questions

1. Using the Periodic Table on page 318 determine the chemical formulas of the following compounds:

   a) sodium chloride

   b) magnesium fluoride

   c) aluminium nitride

   d) lithium oxide

   e) carbon oxide.

2. **SUPPLEMENT** Determine the chemical formulas of the following compounds:

   a) iron(III) oxide (contains Fe^{3+} and O^{2-})

   b) chromium(III) bromide (contains Cr^{3+} and Br^-)

   c) copper(II) sulfate (contains Cu^{2+} and SO_4^{2-}).

3. Balance the following equations.

   a) $Na + Cl_2 \rightarrow NaCl$

   b) $Zn + HCl \rightarrow ZnCl_2 + H_2$

   c) $CuO + C \rightarrow Cu + CO_2$

4. **SUPPLEMENT** Give symbol equations from the following word equations:

   a) carbon + oxygen $\rightarrow$ carbon dioxide

   b) iron + oxygen $\rightarrow$ iron(III) oxide (Fe^{3+} and O^{2-} in iron(III) oxide)

   c) iron(III) oxide + carbon $\rightarrow$ iron + carbon dioxide (Fe^{3+} and O^{2-} in iron(III) oxide)

   d) calcium carbonate + hydrochloric acid $\rightarrow$ calcium chloride + carbon dioxide + water. (CO_3^{2-} is the carbonate ion)

5. Calculate the formula mass of:

   a) ethene, C_2H_4

   b) sulfur dioxide, SO_2

   c) methanol, CH_3OH.

   (A_r: H = 1; C = 12; O = 16: S = 32)

6. Magnesium burns in oxygen to form magnesium oxide:

$2Mg(s) + O_2(g) \rightarrow 2MgO(s)$

$(A_r: O = 16; Mg = 24)$

Calculate:

a) the mass of magnesium required to make 8 g of magnesium oxide

b) the mass of oxygen required to make 8 g of magnesium oxide.

7. Identify the mass of sodium hydroxide that can be made by reacting 2.3 g of sodium with water.

$2Na(s) + 2H_2O(l) \rightarrow 2NaOH(aq) + H_2(g)$

$(A_r: H = 1; O = 16; Na = 23)$

A 8 g

B 4 g

C 12 g

D 6 g

8. **SUPPLEMENT** Calculate how many moles there are in the following.

a) 64 g of S_8

b) 9.8 g of H_2SO_4

c) 21 g of Li

$(A_r: S = 32; H = 1; O = 16; Li = 7)$

9. **SUPPLEMENT** Calculate the mass of the following.

a) 2.5 moles of Sr

b) 0.25 moles of MgO

c) 0.1 moles of C_2H_5Br

$(A_r: Sr = 88; Mg = 24; O = 16; C = 12; H = 1; Br = 80)$

10. SUPPLEMENT Calculate how many moles are in the following.

a) 24 000 cm³ of hydrogen gas, measured at room temperature and pressure.

b) 1200 cm³ of nitrogen gas measured at room temperature and pressure.

11. SUPPLEMENT 0.64 g of copper when heated in air forms 0.80 g of copper oxide.

Deduce the simplest formula of copper oxide.

$(A_r: O = 16; Cu = 64)$

12. SUPPLEMENT Iron(III) oxide is reduced to iron by carbon monoxide.

$(A_r: C = 12; O = 16; Fe = 56)$

$$Fe_2O_3(s) + 3CO(g) \rightarrow 2Fe(s) + 3CO_2(g)$$

a) Calculate the mass of iron that could be obtained by the reduction of 800 tonnes of iron(III) oxide.

b) Calculate the volume of carbon dioxide, measured at room temperature and pressure, that would be obtained by the reduction of 320 g of iron(III) oxide.

13. SUPPLEMENT Determine ionic equations for the following reactions:

a) calcium ions and carbonate ions form calcium carbonate

b) iron(III) ions and hydroxide ions form iron(III) hydroxide

c) silver(I) ions and bromide ions form silver(I) bromide

Practice questions for Sections 1, 2 and 3

Example answer

Note: practice questions, sample answers and comments have been written by the authors. The marks awarded for these questions indicate the level of detail required in the answers. In examinations, the way marks are awarded may be different. References to assessment and/or assessment preparation are the publisher's interpretation of the syllabus requirements and may not fully reflect the approach of Cambridge Assessment International Education.

Question 1

a) The diagrams show the arrangement of particles in the three states of matter.

Each circle represents a particle.

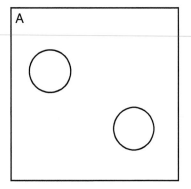 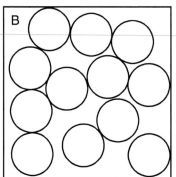 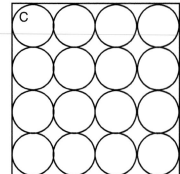

Use the letters A, B, and C to give the starting and finishing states of matter for each of the changes in the table. For the mark, both the starting state and the finishing state need to be correct.

Change	Starting state	Finishing state	
i) The formation of water vapour from a puddle of water on a hot day	B	A	✓ ①
ii) The formation of solid iron from molten iron	B	C	✓ ①
iii) The manufacture of poly(ethene) from ethene	B	A	✗
iv) The reaction whose equation is ammonium chloride(s) → ammonia(g) + hydrogen chloride(g)	C	A	✓ ①

(4)

a) It is important to identify the states of matter:

A = gas, B = liquid, C = solid.

i) Correct – evaporation process.

ii) Correct – solidifying.

iii) Incorrect – order should be 'A, C' because ethene is a gas, poly(ethene) a solid.

iv) Correct – equation shows solid → gases.

b) Answer is 'liquid'. In the Periodic Table at room temperature the majority of elements are solids, a few are gases but only two are liquids – mercury and bromine.

c) i) Correct – sulfur.

   ii) Incorrect – this is a 'mixture' of two elements.

   iii) Correct – a mixture of an element (O_2) and a compound (H_2O).

   iv) Correct – sulfuric acid.

The answers rely on using the state symbols for the equation and a thorough knowledge of the terms elements, mixtures and compounds.

b) Which state of matter is the *least* common for the elements of the Periodic Table at room temperature?

gases ✗   (1)

c) The manufacture of sulfuric acid can be summarised by the equation:

$$2S(s) + 3O_2(g) + 2H_2O(l) \rightarrow 2H_2SO_4(l)$$

Tick one box in each line to show whether the formulas in the table represent a compound, an element or a mixture.

	Compound	Element	Mixture	
i) $2S(s)$		✓		✓ ①
ii) $2S(s)$ $+3O_2(g)$		✓		✗
iii) $3O_2(g) +$ $2H_2O(l)$			✓	✓ ①
iv) $2H_2SO_4(l)$	✓			✓ ①

                                 (4)

(Total 9 marks)

6/9

Question 2

Identify which statement best describes the bonding in water.

A A combination of hydrogen and oxygen molecules.

B Weak bonding within the molecule and strong bonding between molecules.

C Ionic attraction between hydrogen and oxygen ions.

D The sharing of electrons between hydrogen and oxygen atoms.   (1)

(Total 1 mark)

Question 3

This question is about atoms.

a) i) Choose words from the box to label the diagram of an atom.　　(3)

Proton	Neutron	Electron	Ion

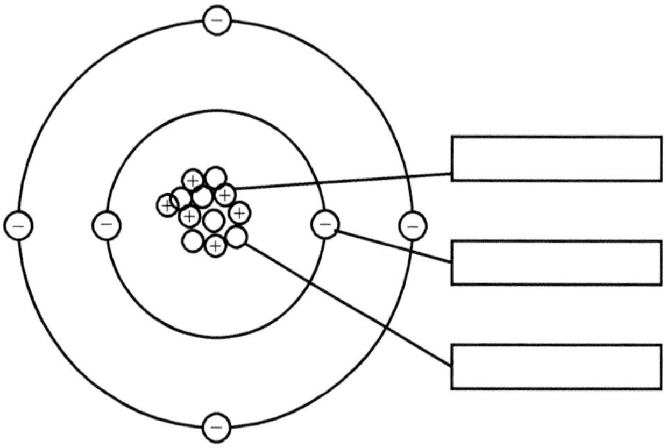

ii) Determine the proton number of this atom.　　(1)

iii) Determine the mass number of this atom.　　(1)

b) Carbon has three isotopes. State one way in which the atoms of the three isotopes are:

i) the same　　(1)

ii) different.　　(1)

(Total 7 marks)

Question 4

a) Some elements combine together to form ionic compounds. Use words from the box to complete the sentences.

Each word may be used once, more than once or not at all.

gained	high	lost	low
medium	metals	non-metals	shared

Ionic compounds are formed between .... and
....

Electrons are .... by atoms of one element and ....
by atoms of the other element.

The ionic compound formed has a .... melting point and a
.... boiling point.　　(6)

b) Two elements react to form an ionic compound with the formula $MgCl_2$.
(proton number of Mg = 12; proton number of Cl = 17)

   i) State the electronic configurations of the two elements in
     this compound *before* the reaction. (2)

   ii) Deduce the electronic configurations of the two elements in
      this compound *after* the reaction. (2)

(Total 10 marks)

Question 5 SUPPLEMENT

Calcium carbonate decomposes on heating, as shown by the equation below.

$$CaCO_3(s) \rightarrow CaO(s) + CO_2(g)$$

Calculate the volume of carbon dioxide produced at room temperature and pressure
when 5 g of calcium carbonate are decomposed. (A_r: C = 12, O = 16, Ca = 40; 1 mole
of gas at room temperature and pressure occupies 24 dm^3)

A 2.4 dm^3

B 4.8 dm^3

C 0.6 dm^3

D 1.2 dm^3 (1)

(Total 1 mark)

Question 6 SUPPLEMENT

9.12 g of iron(II) sulfate was heated. It decomposed to sulfur dioxide ($SO_2(g)$), sulfur
trioxide ($SO_3(g)$) and iron(III) oxide.

Calculate the mass of iron(III) oxide formed and the volume of sulfur trioxide
produced (measured at room temperature and pressure).

$$2FeSO_4(s) \rightarrow Fe_2O_3(s) + SO_2(g) + SO_3(g)$$

(A_r: O = 16, S = 32, Fe = 56; 1 mole of gas at room temperature and pressure occupies
24 000 cm^3) (6)

(Total 6 marks)

Question 7

Consider the structures of the substances shown here:

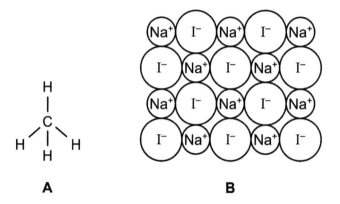

A

B

C

H — Br

D

E

a) Answer these questions using the letters A, B, C, D or E.

i) Identify which structure is methane. (1)

ii) Identify which two structures are giant structures. (1)

iii) Identify which two structures are hydrocarbons. (1)

iv) Identify which structure contains ions. (1)

v) Identify which two structures have very high melting points. (1)

b) Structure E is a form of carbon.

i) State the name of this structure. Put a ring around the correct answer.

carbide   graphite   lead   poly(hexene) (1)

ii) State the name of another form of carbon. (1)

c) Determine the simplest formula for substance B. (1)

d) Is substance D an element or a compound? Explain your answer. (3)

(Total 11 marks)

Question 8 SUPPLEMENT

Strontium and sulfur chlorides both have a formula of the type XCl_2 but they have different properties.

Property	Strontium chloride	Sulfur chloride
Appearance	White crystalline solid	Red liquid
Melting point /°C	873	−80
Particles present	Ions	Molecules
Electrical conductivity of solid	Poor	Poor
Electrical conductivity of liquid	Good	Poor

a) i) Use the Periodic Table to identify which Group strontium is in. (1)

   ii) Which ion does strontium form? (1)

b) Demonstrate the bonding in sulfur chloride using a dot-and-cross diagram.
   Use x to represent an electron from a sulfur atom.
   Use o to represent an electron from a chlorine atom. (3)

c) Explain the difference in electrical conductivity between the following:

   i) solid and liquid strontium chloride (1)

   ii) liquid strontium chloride and liquid sulfur chloride. (1)

(Total 7 marks)

In this section you will explore the chemical reactions that can be caused by using electricity, a process known as electrolysis. You will see that the use of electricity in the development of fuel cells could provide a very effective way of reducing the use of fossil fuels and the associated production of greenhouse gases. There are clear links to Section 2, Ions and ionic bonds, and to Section 9, Metals, which includes the use of electrolysis in the manufacture of aluminium. There are also links with Section 3, Stoichiometry, and the study at supplementary level of the use of ionic half equations.

Starting points

1. Give examples of the common conductors and insulators you are familiar with.

2. Can you list some of the typical properties of ionic and covalent compounds?

3. Have you come across the term *electrolyte*? If so, in what context?

SYLLABUS SECTIONS COVERED

4.1 Electrolysis

4.2 Hydrogen–oxygen fuel cells

4

Electrochemistry

△ Electroplating is used on an industrial scale.

△ Fig. 4.1 Industrial electroplating is a form of electrolysis.

Electrochemistry

INTRODUCTION

Most elements in nature are found combined with other elements as compounds. These compounds must be broken down to obtain the elements that they contain. One of the most efficient and economical ways to break down some compounds is using electricity in a process called electrolysis. Simple electrolysis experiments can be performed in the laboratory, and electrolysis is also used in large-scale industrial processes to produce important chemicals like aluminium and chlorine.

This topic deals with the underlying principles of electrolysis as well as some of the experiments that can be performed in the laboratory and some of the important industrial processes involving electrolysis.

KNOWLEDGE CHECK

✓ There are different arrangements of particles in solids, liquids and gases.
✓ Materials can be classified as 'conductors' or 'insulators'.
✓ In ionic bonding electrons are transferred between atoms, forming ions.
✓ In covalent bonding electrons are shared between atoms, forming molecules.

LEARNING OBJECTIVES

✓ Define electrolysis as the decomposition of an ionic compound, when molten or in aqueous solution, by the passage of an electric current.
✓ Identify in simple electrolytic cells: the anode as the positive electrode; the cathode as the negative electrode; the electrolyte as the molten or aqueous substance that undergoes electrolysis.
✓ Identify the products formed at the electrodes and describe the observations made during the electrolysis of: molten lead(II) bromide; concentrated aqueous sodium chloride; dilute sulfuric acid, using inert electrodes made of platinum or carbon/graphite.
✓ SUPPLEMENT State that metals or hydrogen are formed at the cathode and that non-metals (other than hydrogen) are formed at the anode.
✓ SUPPLEMENT Predict the identity of the products at each electrode for the electrolysis of a binary compound in the molten state.
✓ SUPPLEMENT Describe the transfer of charge during electrolysis to include: the movement of electrons in the external circuit; the loss or gain of electrons at the electrodes; the movement of ions in the electrolyte.
✓ SUPPLEMENT Identify the products formed at the electrodes and describe the observations made during the electrolysis of aqueous copper(II) sulfate using inert carbon/graphite electrodes and when using copper electrodes.
✓ SUPPLEMENT Construct ionic half-equations for reactions at the cathode (showing gain of electrons as a reduction reaction).
✓ State that a hydrogen–oxygen fuel cell uses hydrogen and oxygen to produce electricity with water as the only chemical product.
✓ SUPPLEMENT Describe the advantages and disadvantages of using hydrogen–oxygen fuel cells in comparison with gasoline/petrol engines in vehicles.

ELECTROLYSIS

Compounds that can conduct electricity are called **electrolytes** – they undergo a reaction called **electrolysis**. Electrolysis is defined as the decomposition of an ionic compound, when molten or in aqueous solution, by the passage of an electric current. Experiments can be carried out using a simple electrical **cell**, as shown in Fig. 4.2.

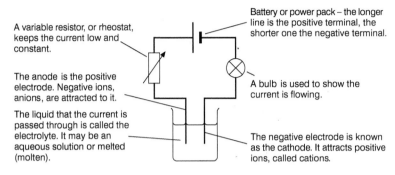

A variable resistor, or rheostat, keeps the current low and constant.

Battery or power pack – the longer line is the positive terminal, the shorter one the negative terminal.

The anode is the positive electrode. Negative ions, anions, are attracted to it.

A bulb is used to show the current is flowing.

The liquid that the current is passed through is called the electrolyte. It may be an aqueous solution or melted (molten).

The negative electrode is known as the cathode. It attracts positive ions, called cations.

Δ Fig. 4.2 A simple electrolysis cell.

When the solution in the beaker is an electrolyte, a complete circuit will form and the bulb will light. The electric current that flows is caused by electrons moving in the electrodes and wires of the circuit, and by ions moving in the solution. If a current does not flow then the beaker must contain a non-electrolyte. Because of this, a simple circuit like this can be used to distinguish between electrolytes and non-electrolytes.

Conditions for electrolysis

The substance being electrolysed (the electrolyte) must contain ions, and these ions must be free to move. In other words, the substance must either be molten or dissolved in water.

A direct current (d.c.) voltage must be used. The **electrode** connected to the positive terminal of the power supply is known as the **anode**. The electrode connected to the negative terminal is known as the **cathode**. The electrical circuit can be drawn as shown in Fig. 4.3.

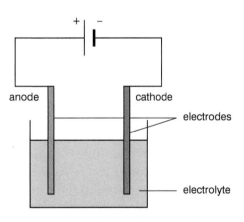

Δ Fig. 4.3 A typical electrical circuit used in electrolysis.

SUPPLEMENT

During electrolysis electrical charge is transferred as follows:

- negative ions move to the anode and give up electrons
- the electrons travel through the anode and around the circuit in the connecting wires to the cathode
- the electrons reaching the cathode are taken up by positive ions.

The reactions at the electrodes can be shown using ionic half-equations. These show the loss of electrons by negative ions at the anode (an example of **oxidation**) and the gain of electrons by positive ions at the cathode (an example of **reduction**).

1. Explain the term *electrolysis*.

2. What is the name given to the positive electrode?

3. Identify the two conditions that must exist for a substance to be an electrolyte and allow an electric current to pass through it.

Electrolysis of molten lead(II) bromide

When an electric current passes through an electrolyte, new substances are formed. The products are lead and bromine as shown in Fig 4.4. The examples below show how you can work out what products will form.

Lead(II) bromide ($PbBr_2$) is ionically bonded and contains Pb^{2+} ions and Br^- ions. When the solid is melted and a voltage is applied, the ions are able to move. The positive lead ions move to the negative electrode (the cathode), and the negative bromide ions move to the positive electrode (the anode). The electrodes are usually made of carbon/graphite, which is inert. This means they do not undergo any **chemical change** during the electrolysis. Silvery deposits of lead form near the bottom of the dish, and brown bromine vapour near the anode.

At the cathode (negative electrode), the lead ions accept electrons to form lead atoms:

$$Pb^{2+}(l) + 2e^- \rightarrow Pb(l)$$

At the anode (positive electrode), the bromide ions give up electrons to form bromine atoms, and then bromine molecules:

$$2Br^-(l) \rightarrow Br_2(g) + 2e^-$$

Note: the two equations above are known as **ionic half-equations**. Unlike normal chemical equations, they do not show the whole chemical change – just the change occurring at an electrode. In the half-equations above, you will see that the numbers of electrons accepted and released are the same. The electric current is produced by this flow of electrons around the external circuit.

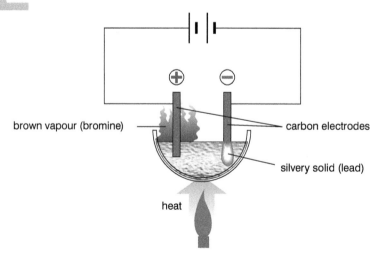

brown vapour (bromine)

carbon electrodes

silvery solid (lead)

heat

Δ Fig. 4.4 Electrolysis of molten lead(II) bromide. This experiment must be carried out by the teacher in a fume cupboard.

The electrolysis of dilute sulfuric acid

When dilute sulfuric acid is electrolysed, hydrogen forms as a colourless gas at the cathode and oxygen as a colourless gas at the anode. In this electrolysis platinum or carbon (graphite) can be used as the electrodes.

SUPPLEMENT

When the sulfuric acid dissolves in water, the hydrogen and sulfate ions separate and are free to move independently. In addition, the water provides a small quantity of hydrogen (H^+) and hydroxide (OH^-) ions:

$$H_2SO_4(aq) \rightarrow 2H^+(aq) + SO_4^{2-}(aq)$$

$$H_2O(l) \rightleftharpoons H^+(aq) + OH^-(aq)$$

The water breaking up to forms ions is an example of **dissociation**. In fact, the ions also combine to form water – the reaction goes both ways: it is a **reversible** reaction. Although there are very few ions present, if they are removed they will be immediately replaced. Therefore, whenever you consider the electrolysis of an aqueous solution you must always include the H^+ and OH^- ions.

When dilute sulfuric acid is electrolysed using inert (graphite) electrodes the following reactions occur at the electrodes.

- At the cathode (negative electrode):
  $H^+(aq)$ ions will be discharged and accept electrons to form hydrogen atoms and then hydrogen molecules (observed as bubbles of a colourless gas)
  $$2H^+(aq) + 2e^- \rightarrow H_2(g)$$

- At the anode (positive electrode):
  The OH^- ions are discharged in preference to the SO_4^{2-} ions to form oxygen (observed as bubbles of a colourless gas)
  $$4OH^-(aq) \rightarrow 2H_2O(l) + O_2(g) + 4e^-$$

Electrolysis of concentrated sodium chloride solution

When concentrated sodium chloride solution is electrolysed, hydrogen ions (from the water solvent) form hydrogen molecules at the cathode and chloride ions form chlorine molecules at the anode. So bubbles of a colourless gas (hydrogen) will be observed at the cathode and bubbles of a pale green gas (chlorine) will be seen at the anode.

This experiment can be performed using a cell as shown in Fig. 4.2. Again, inert electrodes of platinum or carbon (graphite) can be used.

SUPPLEMENT

Electrolysis of copper(II) sulfate solution

Using carbon electrodes

When copper(II) sulfate solution is electrolysed using inert carbon electrodes, the products of the electrolysis are copper and oxygen.

The copper forms as a red-brown coating on the carbon cathode and bubbles of a colourless gas, oxygen, are seen next to the anode.

The solution will become paler blue as the copper ions are discharged.

The following changes occur.

- At the cathode (negative electrode):
  two ions, Cu^{2+} and H^+, move to the cathode and Cu^{2+} ions are discharged.

  $$Cu^{2+}(aq) + 2e^- \rightarrow Cu(s)$$

  The copper ions accept electrons and form copper atoms.

- At the anode (positive electrode):
  two ions, SO_4^{2-} and OH^-, move to the anode and OH^- ions are discharged.

  $$4OH^-(aq) \rightarrow 2H_2O(l) + O_2(g) + 4e^-$$

The hydroxide ions give up electrons and form oxygen molecules.

As the ions are discharged the electrolyte will increasingly contain sulfuric acid (H^+ ions and SO_4^{2-} ions).

Using copper electrodes

When the electrolysis is repeated with copper electrodes, copper is deposited as a red/brown coating at the cathode but there is a difference in the reaction that takes place at the anode.

- At the cathode (negative electrode):
  Cu^{2+} ions gain two electrons and are discharged. Copper atoms are formed and the mass of the electrode increases.

  $$Cu^{2+}(aq) + 2e^- \rightarrow Cu(s)$$

- At the anode (positive electrode):
  copper atoms lose two electrons and Cu^{2+} ions are formed. The anode slowly dissolves and loses mass.

  $$Cu(s) \rightarrow Cu^{2+}(aq) + 2e^-$$

The concentration of the Cu^{2+} ions in the solution remains constant because the rate of production of Cu^{2+} ions at the anode is *exactly balanced* by the rate of removal of Cu^{2+} ions at the cathode. This reaction is important in the refining of copper. Copper is extracted from its ore by reduction with carbon, but the copper produced is not pure enough for many of its uses, such as in electrical cables. It can be purified using electrolysis.

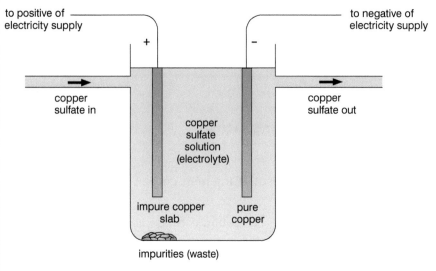

to positive of electricity supply

to negative of electricity supply

+

−

copper sulfate in

copper sulfate out

copper sulfate solution (electrolyte)

impure copper slab

pure copper

impurities (waste)

Δ Fig. 4.5 Copper is purified by electrolysis.

The impure copper is made the anode in a cell with copper(II) sulfate solution as the electrolyte. The cathode is made of a thin piece of pure copper. At the anode the copper atoms form copper ions and impurities fall to the bottom of the tank. The copper ions are then deposited as pure copper on the cathode. The cathode can be replaced by another thin piece of copper once sufficient copper has been deposited.

PREDICTING THE PRODUCTS OF ELECTROLYSIS

Predicting the products of the electrolysis of simple molten ionic compounds is relatively straightforward. The metal forms at the cathode and the non-metal forms at the anode. For example, the electrolysis of molten aluminium oxide forms aluminium (at the cathode) and oxygen (at the anode).

Metals or hydrogen are formed at the cathode and that non- metals (other than hydrogen) are formed at the anode.

QUESTIONS

1. a) Describe what is an *inert* electrode.

   b) Give an example of a substance that is often used as an inert electrode.

2. SUPPLEMENT Deduce what products are formed when the following molten solids are electrolysed.

   a) lead(II) chloride

   b) magnesium oxide

   c) aluminium oxide

3. SUPPLEMENT a) Write a half-equation showing how aluminium ions (Al^{3+}) are discharged.

   b) Suggest at which electrode this change would take place.

Developing practical skills

Fig. 4.6 shows the apparatus that can be used to electrolyse dilute sulfuric acid. (Note: dilute sulfuric acid may cause harm in eyes or a cut.)

Using and organising techniques, apparatus and materials

1. Inert electrodes were used. Suggest what the electrodes were made of.

2. State whether the hydrogen formed at the anode or the cathode.

3. State what name is given to solutions, such as sulfuric acid, that allow an electric current to flow through them.

Observing, measuring and recording

4. Oxygen was collected at the left-hand electrode. How did the volume of oxygen collected compare with the volume of hydrogen collected at the right-hand electrode?

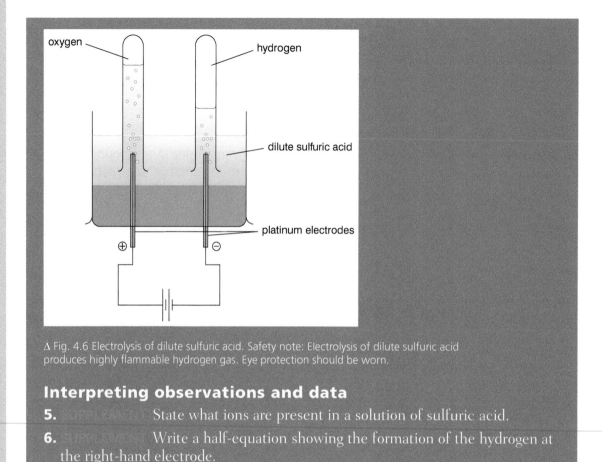

△ Fig. 4.6 Electrolysis of dilute sulfuric acid. Safety note: Electrolysis of dilute sulfuric acid produces highly flammable hydrogen gas. Eye protection should be worn.

Interpreting observations and data

5. State what ions are present in a solution of sulfuric acid.

6. Write a half-equation showing the formation of the hydrogen at the right-hand electrode.

HYDROGEN–OXYGEN FUEL CELLS

Hydrogen and oxygen can be used to produce electricity in a **fuel cell**. The hydrogen gas is pumped onto the anode and oxygen gas onto the cathode. Water is the only substance produced. The platinum catalyst removes an electron from each hydrogen atom (forming hydrogen ions, H^+) and the electrons pass through the external circuit to the anode, producing an electric current. The chemical reactions are complicated but overall the hydrogen and oxygen combine to form water.

△ Fig. 4.7 Hydrogen fuel cell. The platinum, used as the catalyst, is expensive and rare.

SUPPLEMENT

The fact that a hydrogen–oxygen fuel cell only produces water as a by-product is an advantage when compared to the products produced from the combustion of gasoline or petrol. The carbon 'footprint' produced by burning gasoline and petrol has a major negative impact on environmental issues such as global warming. However, most of the hydrogen used today is made by a process which generates carbon dioxide and this is responsible for increasing global warming.

SCIENCE IN CONTEXT

FUEL CELLS

Fuel cells come in many varieties. However, they all work in the same way. They are made up of three segments: the anode, the electrolyte and the cathode. Chemical reactions occur at the interfaces of the three different segments. In a hydrogen–oxygen fuel cell at the anode a catalyst oxidises the hydrogen, producing hydrogen ions and electrons. The electrons travel through a wire creating the electric current. The ions travel through the electrolyte to the cathode where they react with oxygen to form water.

Some hydrogen vehicles burn hydrogen in an internal combustion engine (in the same way petrol or diesel is used). However, it is more common now to react hydrogen with oxygen in a fuel cell and this is a leader in the 'hydrogen economy'. However, as is often the case, the situation is not a simple as you might think. Whilst a hydrogen–oxygen vehicle fuel cell does not produce polluting gases in its exhaust gases, the most common process for the manufacture of hydrogen (producing 98% of the hydrogen used in 2019) produces carbon dioxide.

In addition, fuel cell cars are expensive when compared to petrol or diesel cars.

Δ Fig. 4.8 A hydrogen station.

△ Fig. 4.9 A hydrogen manufacturing plant.

Challenge Question:

a) Write a half-equation showing the formation of hydrogen ions from hydrogen molecules.

b) Write an equation showing the reaction of hydrogen with oxygen.

End of topic checklist

Key terms

anode, cathode, cell, chemical change, electrode, electrolysis, electrolyte, fuel cell, ionic compound

SUPPLEMENT ionic half-equations, oxidation, reduction

During your study of this topic you should have learned:

○ That electrolysis is the breakdown of an ionic compound, molten or in aqueous solution, by the passage of an electric current.

○ How to use the terms anode, cathode and electrolyte.

○ How to describe the electrode products in the electrolysis of these substances, using inert electrodes of platinum or carbon/graphite:

  • molten lead(II) bromide
  • concentrated aqueous sodium chloride
  • dilute sulfuric acid.

○ SUPPLEMENT That metals or hydrogen are formed at the negative electrode (cathode) and that non-metals (other than hydrogen) are formed at the positive electrode (anode).

○ SUPPLEMENT How to predict the products of the electrolysis of a binary (two element) compound in the molten state.

○ That a hydrogen–oxygen fuel cell uses hydrogen and oxygen to produce electricity with water as the only by-product.

○ SUPPLEMENT That the transfer of charge during electrolysis involves the movement of electrons in the external circuit, the loss or gain of electrons at the electrodes and the movement of ions in the electrolyte.

○ SUPPLEMENT Identify the products formed at the electrodes and describe the observations made during the electrolysis of aqueous copper (II) sulfate using carbon/graphite electrodes and when using copper electrodes'

○ SUPPLEMENT How to construct ionic half-equations for reactions at the cathode (reduction).

○ SUPPLEMENT How to describe the advantages and disadvantages of using hydrogen–oxygen fuel cells compared with vehicles powered by gasoline or petrol.

End of topic questions

1. Explain the following terms:

   a) *electrolysis*

   b) *electrolyte*

   c) *electrode*

   d) *anode*

   e) *cathode*.

2. **SUPPLEMENT** When molten lead(II) chloride is electrolysed with graphite electrodes deduce what is observed at each electrode.

	Negative electrode	Positive electrode
A	bubbles of a colourless gas	bubbles of a colourless gas
B	shiny grey solid	bubbles of a pale yellow gas
C	bubbles of a colourless gas	shiny grey solid
D	shiny grey solid	bubbles of a colourless gas

3. **SUPPLEMENT** Zinc bromide, $ZnBr_2$, is an ionic solid. Explain why does the solid not conduct electricity.

4. **SUPPLEMENT** Copy and complete the following table which shows the products formed when molten electrolytes undergo electrolysis.

Electrolyte	Product at the anode	Product at the cathode
Silver bromide		
Lead(II) chloride		
Aluminium oxide		
	Iodine	Magnesium

5. Sodium chloride, NaCl, is an ionic compound. State the products at the anode and cathode in the electrolysis of:

   SUPPLEMENT a) molten sodium chloride

   b) concentrated aqueous sodium chloride.

End of topic questions continued

6. **SUPPLEMENT** Write half-equations for the following reactions:

a) the formation of aluminium atoms from aluminium ions

b) the formation of sodium atoms from sodium ions

7. **SUPPLEMENT** Sodium hydroxide and chlorine are manufactured by the electrolysis of concentrated sodium chloride solution.

a) State at which electrode the chlorine gas is formed.

b) State what gas is formed at the other electrode.

8. **SUPPLEMENT** Describe the advantages and disadvantages of using hydrogen–oxygen fuel cells instead of petrol/gasoline engines in cars.

In this section you will explore some chemical reactions that produce significant amounts of thermal energy, as well as some strange ones that absorb energy and make everything cooler. You will be introduced to energy pathway diagrams which represent the different energy changes that take place in a chemical reaction. There are clear links to Section 2, Molecules and covalent bonds. You will also have plenty of practice at writing chemical equations, linking back to Section 3, Stoichiometry.

Starting points

1. What is a hydrocarbon?

2. How many non-renewable fuels can you name? What products are formed when these fuels burn?

3. Find out what an 'exothermic reaction' is and how this is a key factor in choosing a non-renewable fuel.

SYLLABUS SECTIONS COVERED

5.1 Exothermic and endothermic reactions

5
Chemical energetics

Exothermic and endothermic reactions

△ Fig. 5.1 Fireworks are carefully controlled chemical reactions.

INTRODUCTION

When chemicals react together the reaction causes the transfer of thermal energy to and from the surroundings. This is obvious when a fuel is burned and thermal energy is released into the surroundings. Thermal energy changes in other reactions may be less dramatic but they still take place. A knowledge of chemical bonding can really help to understand how these energy transfers occur.

KNOWLEDGE CHECK

✓ Atoms in molecules are held together by covalent bonds.
✓ Many common fuels are organic compounds called alkanes.
✓ Word and symbol equations show how reactants form products.

LEARNING OBJECTIVES

✓ State that an exothermic reaction transfers thermal energy to the surroundings leading to an increase in the temperature of the surroundings.
✓ State that an endothermic reaction takes in thermal energy from the surroundings leading to a decrease in the temperature of the surroundings.
✓ **SUPPLEMENT** Interpret reaction pathway diagrams showing exothermic and endothermic reactions.
✓ **SUPPLEMENT** State that the transfer of thermal energy during a reaction is called the enthalpy change, ΔH, of the reaction. ΔH is negative for exothermic reactions and positive for endothermic reactions.
✓ **SUPPLEMENT** Define activation energy, E_a, as the minimum energy that colliding particles must have to react.
✓ **SUPPLEMENT** Draw and label reaction pathway diagrams for exothermic and endothermic reactions using information provided, to include: reactants; products; enthalpy change of the reaction, ΔH; activation energy, E_a.
✓ **SUPPLEMENT** State that bond breaking is an endothermic process and bond making is an exothermic process.

THERMAL ENERGY TRANSFER IN CHEMICAL REACTIONS

In most reactions, thermal energy is transferred to the surroundings and the temperature of the surroundings goes up. These reactions are **exothermic**. Some examples of exothermic reactions are combustion and respiration. In a minority of cases, thermal energy is absorbed from the surroundings as a reaction takes place and the temperature of the surroundings goes down. These reactions are **endothermic**. Some examples of endothermic reactions are photosynthesis and thermal decomposition.

For example, when magnesium ribbon is added to dilute hydrochloric acid, the temperature of the acid increases – the reaction is exothermic. In contrast, when sodium hydrogencarbonate is added to hydrochloric acid, the temperature of the acid decreases – the reaction is endothermic.

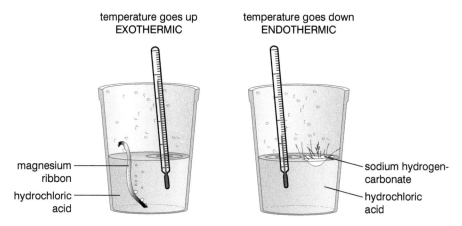

Δ Fig. 5.2 Measuring energy changes in reactions.

Energy changes in reactions like these can be measured using an insulated cup as a calorimeter. If a lid is added to the cup, very little energy is transferred to the air and reasonably accurate results can be obtained.

QUESTIONS

1. Explain an *exothermic* reaction.

2. Explain an *endothermic* reaction.

3. Explain why insulated cups make good calorimeters for measuring energy changes in some chemical reactions.

All reactions involving the **combustion** of fuels are exothermic. The thermal energy transferred when a fuel burns can be measured using **calorimetry**, as shown in the diagram.

The rise in temperature of the water is a measure of the thermal energy transferred to the water. This technique will not give a very accurate answer because much of the energy will be transferred to the surrounding air. Nevertheless, it can be used to compare the energy released by **burning** the same amounts of different fuels.

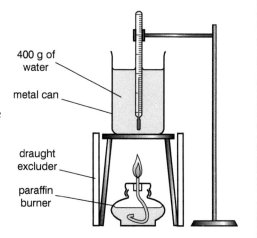

Δ Fig. 5.3. Measuring the thermal energy transferred by by burning a liquid fuel.

1. Measuring energy changes when burning fuels in a liquid burner does not give very accurate results. Explain why this is.

2. A group of students was comparing the energy released on burning different liquid fuels in spirit burners using apparatus similar to that shown in Fig. 5.3. The results obtained are shown here:

Name of fuel	Amount of fuel burned (g)	Rise in temperature of 200 cm³ of water in a metal can (°C)	Temperature rise of the water per g of fuel burned (°C/g)
Ethanol	1.1	32	
Paraffin	0.9	30	
Pentane	1.5	38	
Octane	0.5	20	

   a) Suggest how the students worked out how much fuel was burned in each experiment.

   b) Complete the last column of the table by calculating the temperature rise in each experiment per gram of fuel burned. (Give your answers to the nearest whole number.)

   c) i) Suggest which fuel produced the greatest temperature rise per gram burned.

   ii) If the octane experiment were repeated using 400 cm³ of water in the metal can, calculate what temperature rise would you expect. Explain how you worked out your answer.

3. Another group of students used a glass beaker rather than a metal can in their experiments. Suggest which group of students would get more accurate results. Explain your answer.

Reaction pathway diagrams

Reaction pathway diagrams are used to show the overall energy changes in reactions. Some reaction pathway diagrams are shown in Fig. 5.4. They show the relative energy levels of the reactants and the products. In an exothermic reaction the energy level of the products is lower than that of the reactants. In an endothermic reaction the energy level of the products is higher than that of the reactants.

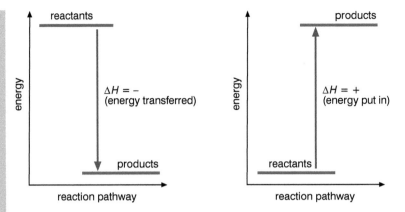

△ Fig. 5.4 Reaction pathway diagrams for exothermic and endothermic reactions.

The transfer of thermal energy in chemical reactions is called the **enthalpy change**. It is given the symbol ΔH. The enthalpy change for a particular reaction is shown at the end of the balanced equation. The units are kJ/mol.

Reaction pathway diagrams show the enthalpy difference between the reactants and the products.

In an exothermic reaction, the energy content of the reactants is greater than the energy content of the products. Energy is being transferred to the surroundings. ΔH is negative.	In an endothermic reaction, the energy content of the products is greater than the energy content of the reactants. Energy is being absorbed from the surroundings. ΔH is positive.

All ΔH values should have a + or − sign in front of them to show if they are endothermic or exothermic.

Activation energy

Activation energy (E_a) is the *minimum* amount of energy that colliding particles must have in order to react. Fig. 5.5a shows the activation energy of an exothermic reaction and Fig. 5.5b shows the activation energy of an endothermic reaction. A reaction will only take place if the necessary activation energy is available.

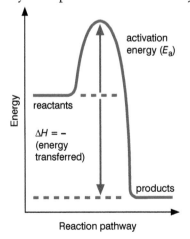

△ Fig. 5.5a Reaction pathway diagram (energy profile) for an exothermic reaction.

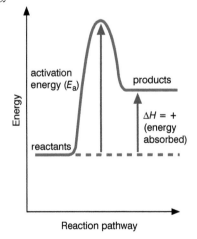

△ Fig. 5.5b Reaction pathway diagram for an endothermic reaction.

1. SUPPLEMENT The overall energy change for a particular reaction is positive. Is the reaction endothermic or exothermic?

2. SUPPLEMENT In an endothermic reaction how does the energy released in the reaction compare to the activation energy?

3. SUPPLEMENT On a reaction pathway diagram, suggest the name given to the minimum amount of energy required for a reaction to occur.

4. SUPPLEMENT Methane burns in excess oxygen to form carbon dioxide and water. When 1 mole of methane is burnt 882 kJ of energy is released. Draw a labelled energy profile/pathway diagram for this reaction.

5. SUPPLEMENT Draw and label energy profile/pathway diagrams for each of the reactions shown in the table below:

Reaction	Activation energy (kJ)	Overall energy change (kJ)
A	120	−90
B	80	+10

Developing practical skills

Two students used an insulated cup to compare the energy changes in two reactions:

• magnesium with hydrochloric acid
• sodium hydrogencarbonate with hydrochloric acid.

They added 50 cm³ of 0.5 mol/dm³ hydrochloric acid to an insulated cup and measured its temperature. They then added a known mass of magnesium ribbon to the acid. They stirred the reaction mixture with a glass stirring rod until the reaction was complete and then took the final temperature. They then repeated the procedure using sodium hydrogencarbonate. Their results are shown in the table.

Reaction	Mass of solid used (g)	Initial temperature of the acid (°C)	Final temperature of the acid (°C)	Temperature change (°C)
Magnesium + hydrochloric acid	0.1	22	35	
Sodium hydrogencarbonate + hydrochloric acid	2.5	22	18	

Using and organising techniques, apparatus and materials

1. Suggest what apparatus was used to measure the volumes of hydrochloric acid.

2. Explain why the students stir the reaction mixtures until the reaction was complete and the final temperature was taken.

Observing, measuring and recording

3. Predict what you would observe when magnesium ribbon is added to dilute hydrochloric acid.

Interpreting observations and data

4. Calculate the temperature change in each reaction.

5. Calculate the temperature change per gram of solid in each reaction.

6. In each case state if the reaction is exothermic or endothermic.

7.   Draw a simple reaction pathway diagram to represent the reaction between magnesium and hydrochloric acid.

Evaluating methods

8. Energy is often transferred from the reaction mixture to the atmosphere or transferred to the reaction mixture from the atmosphere. Suggest one way this error could be reduced.

Where does the energy come from?

The reaction that occurs when a fuel is burning can be considered to take place in two stages. In the first stage the covalent bonds between the atoms in the fuel molecules and between the atoms in the oxygen molecules are broken. In the second stage the atoms combine and new covalent bonds are formed in the products. For example, in the combustion of propane:

propane + oxygen → carbon dioxide + water

$$C_3H_8(g) + 5O_2(g) \rightarrow 3CO_2(g) + 4H_2O(l)$$

$$\Delta H = -2202 \text{ kJ/mol}$$

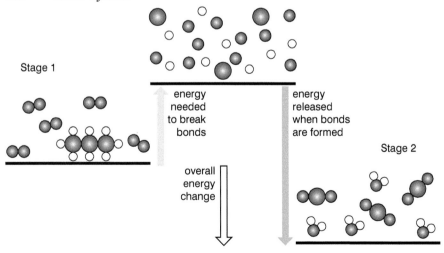

△ Fig. 5.6 Energy changes in an exothermic reaction.

Stage 1: Energy is needed (absorbed from the surroundings) to break the bonds in the reactants. This process is endothermic.

Stage 2: Energy is released (transferred to the surroundings) as the new bonds form in the products. This process is exothermic.

The overall reaction is exothermic because forming the new bonds releases more thermal energy than is needed initially to break the bonds in the reactants. Fig. 5.7 is a simplified reaction pathway diagram showing the exothermic nature of the reaction.

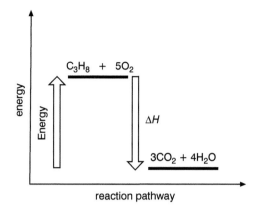

△ Fig. 5.7 A simplified reaction pathway diagram for the reaction.

REMEMBER

In an exothermic reaction, the energy released on forming new bonds in the products is greater than breaking the bonds in the reactants.

In an endothermic reaction, more energy is needed to break the bonds in the reactants than is released when the bonds in the products are formed. The energy changes in endothermic reactions are usually relatively small.

QUESTIONS

1. SUPPLEMENT What does the sign of ΔH indicate about a reaction?

2. SUPPLEMENT Suggest whether energy is needed or released when bonds are broken.

3. SUPPLEMENT In an endothermic reaction suggest whether more or less energy is needed to break the old bonds than is recovered when new bonds are formed.

SCIENCE IN CONTEXT   HOW COMMON ARE ENDOTHERMIC REACTIONS?

Almost all chemical reactions in which simple compounds or elements react to make new compounds are exothermic. One exception is the formation of nitrogen oxide (NO) from nitrogen and oxygen. Overall, energy is needed to create this compound, with less energy being released on forming bonds than was needed to break the bonds of the reactants. Nitrogen oxide is often formed in lightning storms. The lightning provides enough energy to split the nitrogen and oxygen molecules before the atoms combine to form nitrogen oxide:

$N_2(g) + O_2(g) \rightarrow 2NO(g)$ ΔH positive

Another exception is **photosynthesis**. Plants use energy from sunlight to convert carbon dioxide and water into glucose and oxygen:

$$6CO_2(g) + 6H_2O(l) \rightarrow C_6H_{12}O_6(aq) + 6O_2(g) \; \Delta H \text{ positive}$$

△ Fig. 5.8 These plants are making food by photosynthesis, an endothermic reaction.

'Cold packs', which you can buy in some countries, can be used to help you keep cool. Usually you have to bend a pack to break a partition inside and allow two substances to mix. The pack will then stay cold for an hour or longer. However, it may not be an endothermic reaction that is working in the cold pack. Dissolving chemicals like urea or ammonium nitrate in water also cause the temperature of water to fall, but dissolving is a **physical change**, not a chemical change. Whether it is an endothermic reaction or not is the manufacturer's secret.

Challenge Question: Draw a reaction pathway diagram for the photosynthesis equation shown above.

△ Fig. 5.9 A cold pack.

End of topic checklist

burning, calorimetry, combustion, endothermic, exothermic, photosynthesis, physical change

SUPPLEMENT activation energy, enthalpy change, ΔH

During your study of this topic you should have learned:

○ That an exothermic reaction transfers thermal energy to the surroundings, leading to an increase in the temperature of the surroundings.

○ That an endothermic reaction takes in thermal energy from the surroundings, leading to a decrease in the temperature of the surroundings.

○ **SUPPLEMENT** How to interpret reaction pathway diagrams showing exothermic and endothermic reactions.

○ **SUPPLEMENT** That enthalpy change, ΔH, is the transfer of thermal energy during a reaction (ΔH is negative for an exothermic reaction and positive for an endothermic reaction).

○ **SUPPLEMENT** That activation energy, E_a, is the minimum energy that colliding particles must have in order to react.

○ **SUPPLEMENT** How to draw and label reaction pathway diagrams for exothermic and endothermic reactions.

○ **SUPPLEMENT** That bond breaking is an endothermic process and bond making is an exothermic process.

1. State which of the following statements is correct for an endothermic reaction.

   A The temperature of the reaction mixture increases.

   B Thermal energy is absorbed from the surroundings.

   C Thermal energy is released to the surroundings.

   D Most everyday reactions are endothermic reactions.

2. Explain each of the following.

   a) An insulated cup is used when measuring energy changes in simple reactions, such as adding magnesium ribbon to an acid.

   b) When sodium hydrogencarbonate is added to a solution of an acid the temperature of the acid falls.

3. An estimate of the energy transferred by the reaction when a fuel burns can be made by burning the fuel under a container holding water and measuring the temperature rise of the water.

   a) State what type of material the container should be made of. Explain your answer.

   b) Explain why this method gives an estimate rather than an accurate value.

   c) Explain how can the accuracy of this method could be improved.

4. **SUPPLEMENT** Calcium oxide reacts with water as shown in the equation:

   $$CaO(s) + H_2O(l) \rightarrow Ca(OH)_2(s)$$

   A reaction pathway diagram for this reaction is shown below.

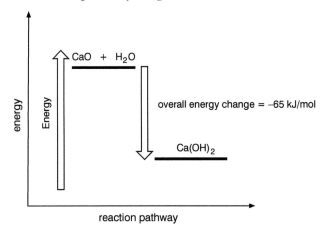

a) State what the reaction pathway diagram tells us about the type of overall energy change that takes place in this reaction.

b) Explain what the reaction pathway diagram indicates about the amount of energy required to break the original bonds and form new bonds in this reaction.

5. **SUPPLEMENT** Draw a reaction pathway diagram for an exothermic reaction. Label the activation energy and the enthalpy change of the reaction.

6. **SUPPLEMENT** Chlorine (Cl_2) and hydrogen (H_2) react together to make hydrogen chloride (HCl). The equation can be written as:

H–H + Cl–Cl → H–Cl + H–Cl

When this reaction occurs, energy is transferred to the surroundings. Explain this in terms of the energy transfer processes taking place when bonds are broken and when bonds are made.

In this section you will explore the speed or rate of chemical reactions as well as how the rate can be controlled. You will also learn about redox reactions which, as the name suggests, involve both reduction and oxidation. You will also have opportunities to practise writing and balancing chemical equations, as introduced in Section 3, Stoichiometry.

Starting points

1. Give an example of a very rapid, almost instantaneous, chemical reaction. Now give an example of a very slow one.

2. What is a catalyst? Can you give any examples of catalysts used in everyday life?

SYLLABUS SECTIONS COVERED

6.1 Physical and chemical changes

6.2 Rate of reaction

6.3 Redox

6
Chemical reactions

△ Rusting is a very slow reaction.

Rate of reaction

Δ Fig. 6.1 Petrol igniting.

INTRODUCTION

Some chemical reactions take place extremely quickly. For example, when petrol is ignited it combines with oxygen almost instantaneously. Reactions like these have a *high rate*. Other reactions are much slower, for example when an iron bar rusts in the air; reactions like these have a *low rate*. Chemical reactions can be controlled and made to be quicker or slower. This can be very important in situations like food production, either by slowing down or increasing the rate at which food ripens, or in the chemical industry, where the rate of a reaction can be adjusted to an optimum level.

KNOWLEDGE CHECK

✓ The particles in the three states of matter (solid, liquid and gas) have different arrangements, movement and energy.
✓ The course of a reaction can be shown in a reaction pathway diagram.
✓ Balanced chemical equations are used to describe reactions.

LEARNING OBJECTIVES

✓ Identify physical and chemical changes, and describe the differences between them.
✓ Describe the effect on the rate of reaction of changing: the concentration of solutions, the pressure of gases, the surface area of solids, the temperature, and adding or removing a catalyst.
✓ State that a catalyst increases the rate of a reaction and is unchanged at the end of a reaction.
✓ Describe practical methods for investigating the rate of a reaction including change in mass of a reactant or a product and the formation of a gas.
✓ Interpret data, including graphs, from rate of reaction experiments.
✓ SUPPLEMENT Describe collision theory in terms of: number of particles per unit volume; frequency of collisions between particles; kinetic energy of particles; activation energy, E_a.
✓ SUPPLEMENT Explain the effect on the rate of reaction of changing: the concentration of solutions, the pressure of gases, the surface area of solids, the temperature, and adding or removing a catalyst, using collision theory.
✓ SUPPLEMENT State that a catalyst decreases the activation energy, E_a, of a reaction.

PHYSICAL AND CHEMICAL CHANGES

A **chemical change**, or chemical reaction, is quite different from **physical changes** that occur, for example, when sugar dissolves in water.

In a chemical change, one or more new substances are produced. In many cases an observable change is apparent, for example, the colour changes or a gas is produced.

An apparent change in mass can occur. This change is often quite small and difficult to detect unless accurate balances are used. Mass is conserved in *all* chemical reactions – the apparent change in mass usually occurs because one of the reactants or products is a gas (whose mass may not have been measured).

An energy change is almost always involved. In most cases energy is released and the surroundings become warmer. In some cases energy is absorbed from the surroundings, and so the surroundings become colder. Note: Some physical changes, such as evaporation, also have energy changes.

RATE OF REACTION

A quick reaction takes place in a short time. It has a high **rate of reaction**. As the time taken for a reaction to be completed increases, the rate of the reaction decreases. In other words:

Speed	Rate	Completion time
Quick or fast	High	Short
Slow	Low	Long

△ Table 6.1 Speed, rate and time.

SUPPLEMENT

Collision theory

For a chemical reaction to occur, the reacting particles (atoms, molecules or ions) must collide. The energy involved in the collision must be enough to break the chemical bonds in the reacting particles – or the particles will just bounce off one another. The more particles there are in a particular volume, the greater the frequency of collisions. Similarly the greater the kinetic energy of the particles, the greater the frequency of collisions.

A collision that has enough energy to result in a chemical reaction is an **effective collision**.

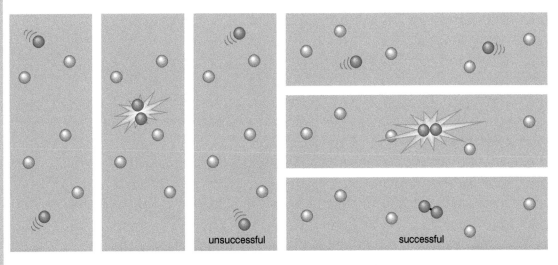

unsuccessful   successful

△ Fig. 6.2 Particles must collide with sufficient energy to make an effective collision.

Some chemical reactions occur extremely quickly (for example, the explosive reaction between petrol and oxygen in a car engine) and some more slowly (for example, iron rusts over days or weeks). This is because they have different **activation energies**. Activation energy acts as a barrier to a reaction. It is the minimum amount of energy required in a collision for a reaction to occur. As a general rule, the bigger the activation energy, the slower the reaction will be at a particular temperature.

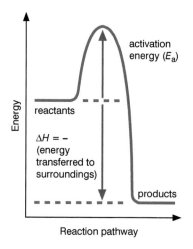

△ Fig. 6.3 Reaction pathway diagram.

REMEMBER

The 'barrier' preventing a reaction from occurring is called the activation energy. If the activation energy of a reaction is low, more of the collisions will be effective and the reaction will proceed quickly. If the activation energy is high, a smaller proportion of collisions will be effective and the reaction will be slower.

QUESTIONS

1. Suggest the main difference between a physical change and a chemical change.

2. In a chemical change there is often an apparent change in mass even though mass cannot be created or destroyed in a chemical reaction. Suggest a possible cause of this apparent change in mass.

3. SUPPLEMENT In the collision theory, what two things must happen for two particles to react?

4. SUPPLEMENT Describe an *effective* collision.

5. SUPPLEMENT Describe, using a diagram, what is meant by the term *activation energy*.

Monitoring the rate of a reaction

The rate of a reaction changes as the reaction proceeds. There are some easy ways of monitoring this change.

When marble (calcium carbonate) reacts with hydrochloric acid, the following reaction starts straight away:

calcium carbonate + hydrochloric acid → calcium chloride + carbon dioxide + water

$$CaCO_3(s) + 2HCl(aq) \rightarrow CaCl_2(aq) + CO_2(g) + H_2O(l)$$

The reaction can be monitored as it proceeds either by measuring the volume of gas being formed or by measuring the change in mass of the reaction flask.

The volume of gas produced in this reaction can be measured using the apparatus shown in Fig. 6.4. The hydrochloric acid is put into the conical flask, the marble chips are added, the stopper is quickly fixed into the neck of the flask and the stopclock is started.

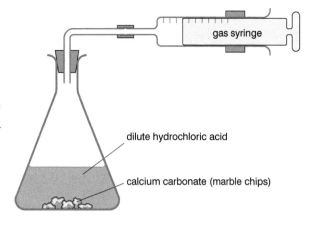

The reaction will start immediately, effervescence (bubbling) will occur in the flask as the carbon dioxide gas is produced and the plunger on the syringe will start to move. Measuring the volume of gas in the syringe every

Δ Fig. 6.4 Monitoring the rate of a reaction.

10 seconds will indicate how the total amount of gas produced changes as the reaction proceeds. The change in the rate of the reaction with time can be shown on a graph of the results (see Fig. 6.6).

To measure the change in mass in the same reaction, the apparatus shown in Fig. 6.5 can be used. The hydrochloric acid is put into the conical flask, the marble chips are added, the cotton wool plug is put in the neck of the flask and the stopclock is started. The mass of the flask and contents is measured as soon as the plug is inserted and then every 10 seconds as the reaction occurs. The mass will decrease as carbon dioxide gas escapes from the flask.

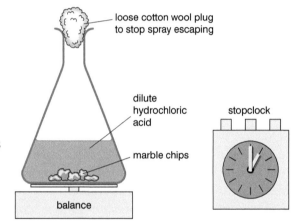

Drawing a graph of the results shows the change in the rate of the reaction over time.

Δ Fig. 6.5 Measuring the change in mass.

Graphs of the results from both experiments have almost identical shapes. The rate of the reaction decreases as the reaction proceeds.

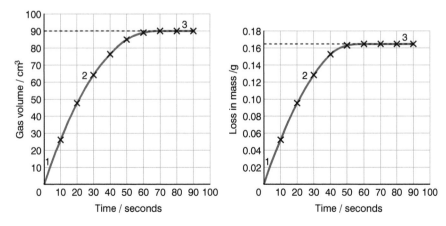

Δ Fig. 6.6 Volume of carbon dioxide produced or loss in mass.

Interpreting graphs from rate of reaction experiments

The rate of the reaction at any point can be calculated from the gradient of the curve. The shapes of the graphs can be divided into three regions.

1. At this point, the curve is the steepest (has the highest gradient) and the reaction has its highest rate. The maximum number of reacting particles are present and the number of effective collisions per second is at its greatest.

2. The curve is not as steep (has a lower gradient) at this point and the rate of the reaction is lower. Fewer reacting particles are present and so the number of effective collisions per second is less.

3. The curve is horizontal (gradient is zero) and the reaction is complete. At least one of the reactants has been completely used up and so no further collisions can occur between the two reactants.

REMEMBER

In experiments like these it is helpful to have a good understanding of the types of **variables** involved. The factor you are investigating is called the **independent variable** – when investigating how the reaction between marble and hydrochloric acid changes over time, time is the independent variable. A **dependent variable** is changed by the independent variable – in the marble and hydrochloric acid reaction, the volume of carbon dioxide produced is the dependent variable. Other variables involved are **control variables** and are not allowed to change to ensure a 'fair test'. So temperature could be a control variable in the reaction between marble and hydrochloric acid.

In chemical reactions it is very rare that exact (as predicted by the equation) quantities of reactants are used. In the marble and hydrochloric acid reaction all the marble may be used up (it is called the *limiting reactant*) but not all the hydrochloric acid; some is left when the reaction has stopped (it is *in excess*).

QUESTIONS

1. What piece of apparatus can accurately measure the volume of gas produced in a reaction?

2. On a volume versus time graph (as in Fig 6.6), explain what a horizontal line shows.

3. When comparing two reactions, on the same graph, will the slower or quicker reaction have a steeper volume/time gradient at the beginning?

What can change the rate of a reaction?

There are five key factors that can change the rate of a reaction:

- concentration (of a solution)
- pressure (of a gas)
- temperature
- surface area/particle size (of a solid)
- a catalyst.

Concentration

Increasing the concentration of a reactant will increase the rate of reaction. When a piece of magnesium ribbon is added to a solution of hydrochloric acid, the following reaction occurs:

magnesium + hydrochloric acid → magnesium chloride + hydrogen

$Mg(s)$ + $2HCl(aq)$ → $MgCl_2(aq)$ + $H_2(g)$

As the magnesium and acid come into contact, there is effervescence ('bubbling') and hydrogen gas is given off. Two experiments were performed using the same length of magnesium ribbon, but different concentrations of acid. In experiment 1 the hydrochloric acid used was 2.0 mol/dm³, in experiment 2 the acid was 0.5 mol/dm³. The graph in Fig. 6.7 shows the results of the two experiments.

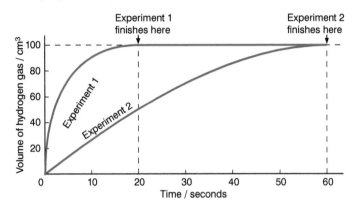

Δ Fig. 6.7 Volume of hydrogen produced in the reaction between magnesium and hydrochloric acid.

In experiment 1 the curve is steeper (has a higher gradient) than in experiment 2. In experiment 1 the reaction is complete after 20 seconds, whereas in experiment 2 it takes 60 seconds. The initial rate of the reaction is higher with 2.0 mol/dm³ hydrochloric acid than with 0.5 mol/dm³ hydrochloric acid.

SUPPLEMENT

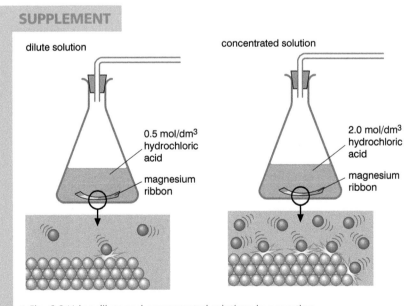

Δ Fig. 6.8 Using dilute and concentrated solutions in a reaction.

Developing practical skills

Two students were investigating how the concentration of hydrochloric acid affects the rate of its reaction with calcium carbonate (marble chips). The reaction produces carbon dioxide gas and the time it takes to produce a certain volume of gas can be used as a measure of the rate of the reaction. They set up the apparatus shown in the diagram.

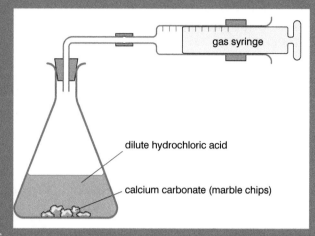

△ Fig. 6.9 Experiment with hydrochloric acid and calcium carbonate. Safety note: Eye protection should be worn and great care taken in supporting and using the gas syringe, especially if the plunger may be pushed out completely.

They placed 40 cm³ of hydrochloric acid in the conical flask and added 10 cm³ of distilled water. One student then carefully weighed 5 g of calcium carbonate (an excess) into a small beaker and then added the calcium carbonate to the flask. As soon as the marble chips were added the other student attached the delivery tube to the gas syringe and started the stop-watch. They then measured the time it took for 40 cm³ of gas to be produced. They repeated the process using the quantities of hydrochloric acid and water shown in the table. The results are shown in the table.

Volume of hydrochloric acid (cm³)	40	35	30	25	20
Volume of water (cm³)	10	15	20	25	30
Time taken for 40 cm³ of gas to be produced (s)	14	18	23	36	67

Using and organising techniques apparatus, and materials

1. Explain why the total volume in the flask was always 50 cm³.

2. Suggest the name of the apparatus you would use to measure the volumes of hydrochloric acid and water.

Interpreting observations and data

3. Draw a graph of volume of hydrochloric acid against time taken to produce 40 cm³ of gas. Draw a smooth curve through the points.

4. Explain what the shape of the curve tells you about the effect of changing the concentration of hydrochloric acid on the rate of the reaction.

5. Use the results to predict what the time would have been if 22 cm³ of hydrochloric acid and 28 cm³ water had been used.

Evaluating methods

6. Evaluate the main sources of error in the design of this experiment.

Temperature

Increasing the temperature of the reactants will increase the rate of a reaction.

SUPPLEMENT

Warming a substance transfers kinetic energy to its particles. More kinetic energy means that the particles move faster. Because they are moving faster there will be more collisions each second. The increased energy of the collisions also means that the proportion of collisions that are effective will increase. A reaction was carried out at two different temperatures – first at 20 °C and then at 30 °C.

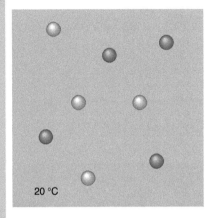
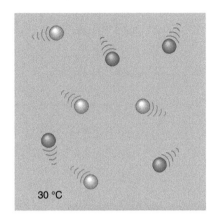

20 °C   30 °C

Δ Fig. 6.10 Effect of increasing temperature on particles.

Increasing the temperature of the reaction between some marble chips and hydrochloric acid will not increase the final amount of carbon dioxide produced. The same amount of gas will be produced in a shorter time. The rates of the two reactions are different but the final loss in mass is the same.

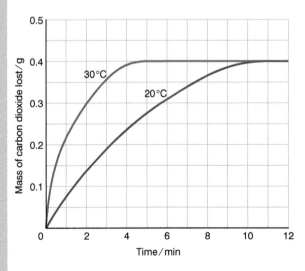

Δ Fig. 6.11 The effect of temperature on the reaction between hydrochloric acid and marble chips.

1. **SUPPLEMENT** In terms of particles colliding, explain why increasing the concentration of a solution increases the rate of reaction.

2. Plan and design an experiment to compare the effect of temperature on the rate of reaction between magnesium and hydrochloric acid. You should:

   a) demonstrate how to safely use techniques, apparatus and materials

   b) plan the procedure you will use

   c) show how you will record your measurements or observations.

3. **SUPPLEMENT** Explain two reasons why increasing temperature increases the rate of reaction.

SCIENCE IN CONTEXT

THE EXPLOSIVE TRUTH ABOUT FLOUR MILLS

The surface area of particles really does affect the rate of some reactions!

Baking bread is a common and important activity, but making the flour that goes into the bread can be a dangerous business. Ever since a serious explosion at a flour mill near Minneapolis in the USA in 1878 killed 18 people, the milling industry has tried to reduce the risk of flour particles igniting into 'flour bombs'. In fact flour dust is thought to be more explosive than coal dust! Similar explosions have occurred in other factories when dust has exploded.

The key components to a flour or dust explosion are very small particles suspended in a plentiful supply of air, in a confined space and with a source of ignition. In factories the source of ignition doesn't have to be something obvious such as a discarded cigarette or match; it could be a spark from an electric motor or other electrical device, or even a light switch. In the case of the 1878 flour mill explosion, the cause of the explosion was thought to be a spark from an ageing electric motor.

△ Fig. 6.12 Dropping milk powder on a flame.

In the laboratory or at home, if you put a match to some flour you might be able to get it to burn, but it certainly won't explode – the flour needs to be suspended in the air as very small particles that are close enough together so that if one flour particle ignites it starts a rapid chain reaction with other particles and then an explosion. So don't under-estimate the importance of particle size on the rate of some reactions.

Challenge Question: In all sorts of underground mines there will be dust. In a coal mine there is an additional hazard that can cause an explosion. Suggest another source of an explosion and explain why you might find it in a coal mine.

Surface area

Increasing the **surface area** (or decreasing the particle size) of a solid reactant will increase the rate of a reaction.

SUPPLEMENT

A reaction can only take place if the reacting particles collide. This means that the reaction takes place at the surface of a solid. The particles within the solid cannot react until those on the surface have reacted and moved away.

Powdered calcium carbonate has a smaller particle size (or much larger surface area) than the same mass of marble chips. A lump of coal will burn slowly in the air, whereas coal dust can react explosively. This is a hazard in coal mines where coal dust can react explosively with air. In addition, as well as the danger of explosive mixtures of coal dust and air, the build-up of methane gas can also form an explosive mixture with the air.

Δ Fig. 6.13 Powdered carbon has a much larger surface area than the same mass in larger lumps.

Catalysts

A **catalyst** is a substance that alters the rate of a chemical reaction and is chemically unchanged at the end of the reaction.

SUPPLEMENT

Most catalysts work by providing an alternative 'pathway' for the reaction – one that has a lower activation energy. The lower activation energy means that more of the collisions between particles will be effective.

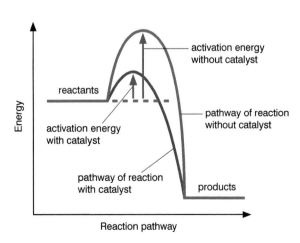

Δ Fig. 6.14 The catalyst provides a lower energy route from reactants to products.

One example of the effect of a catalyst on a reaction is the use of manganese(IV) oxide in the decomposition of hydrogen peroxide. Hydrogen peroxide decomposes at room temperature into water and oxygen. The rate of this reaction is considerably increased by adding manganese(IV) oxide. As a gas is produced, the rate of the reaction can be monitored by collecting the gas in a gas syringe.

$$2H_2O_2(aq) \rightarrow 2H_2O(l) + O_2(g)$$

Catalysts are often used in industry to manufacture important chemicals. Table 6.2 shows some important industrial catalysts.

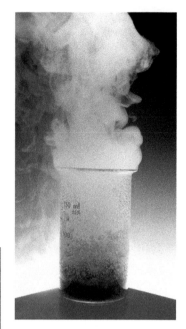

△ Fig. 6.15 The effect of manganese(IV) oxide catalyst on the decomposition of hydrogen peroxide.

Industrial process	Catalyst used
Manufacture of ethanol	Phosphoric acid
Cracking long-chain alkanes	Silica or alumina
Manufacture of ammonia	Iron
Manufacture of sulfuric acid	Vanadium(V) oxide

△ Table 6.2 Uses of catalysts.

Pressure

Increasing the pressure on the reaction involving a gas can increase the rate of reaction. For example, in the reaction between carbon monoxide and oxygen, carbon dioxide is formed as shown in the equation:

$$2CO(g) \ + \ O_2(g) \ \rightarrow \ 2CO_2(g)$$

The reacting gases will react at a greater rate if the pressure is increased.

SUPPLEMENT

The reason for this increase in reaction rate can be explained in terms of **collision theory**. As the pressure is increased the volume of the gases will reduce and so there will be a greater number of particles per unit volume. As a result there will be a greater frequency of collisions between the particles. This will lead to more effective collisions taking place in a given time, so that there will be more collisions with sufficient energy to exceed the activation energy needed for the reaction to take place.

QUESTIONS

1. Describe a *catalyst*.

2. Suggest in what type of reactions pressure is likely to have a significant effect on the rate of the reaction.

3. SUPPLEMENT Use collision theory to explain why changing the surface area of a solid in a reaction with an acid will affect the rate of that reaction.

End of topic checklist

catalyst, chemical change, control variable, dependent variable, independent variable, physical change, rate of reaction, surface area, variable

SUPPLEMENT activation energy, collision theory, effective collision, reaction pathway diagram

During your study of this topic you should have learned:

○ How to identify physical and chemical changes and understand the differences between them.

○ How to describe the effect of concentration, pressure of gases, surface area, catalysts and temperature on the rate of reactions.

○ That a catalyst increases the rate of a reaction and is unchanged at the end of the reaction.

○ How to describe a practical method for investigating the rate of a reaction involving the change in mass of reactants or products and evolution of a gas.

○ How to interpret data, including graphs, from rate of reaction experiments.

○ **SUPPLEMENT** How to describe collision theory when referring to the number of particles per unit volume, the frequency of collisions between particles, the kinetic energy of the particles and the activation energy.

○ **SUPPLEMENT** How to describe and explain the effects of concentration, surface area, temperature, catalysts and pressure of gases on reaction rate.

○ **SUPPLEMENT** That a catalyst decreases the activation energy of a reaction.

1. Select which of the following changes will NOT significantly affect the rate of the reaction between sodium thiosulfate solution and dilute hydrochloric acid:

   A raising the temperature of the hydrochloric acid.

   B increasing the concentration of the sodium thiosulfate solution.

   C decreasing the concentration of the hydrochloric acid.

   D increasing the pressure of the air in the room where the reaction takes place.

2. This question is about the reaction between magnesium and hydrochloric acid.

   a) Draw and label a diagram of the apparatus that could be used to monitor the rate of the reaction by measuring the volume of hydrogen produced.

   b) Predict how the following changes affect the rate of the reaction:

      i) using powdered magnesium rather than magnesium ribbon

      ii) using a less concentrated solution of hydrochloric acid

      iii) lowering the temperature of the hydrochloric acid.

3. Explain why there is a risk of an explosion in a flour mill.

4. Look at the table of results obtained when dilute hydrochloric acid is added to marble chips.

Time (seconds)	0	10	20	30	40	50	60	70	80	90
Volume of gas (cm³)	0	20	36	49	58	65	69	70	70	70

   a) State the name of the gas produced in this reaction.

   b) Write a word equation, including state symbols, for the reaction.

   c) Draw a graph of volume of gas (y-axis) against time (x-axis).

   Label it 'Graph 1'.

   d) Use the results to calculate the volume of gas produced:

      i) in the first 10 seconds

      ii) between 10 and 20 seconds

      iii) between 20 and 30 seconds

      iv) between 80 and 90 seconds.

   e) SUPPLEMENT Explain why the rate of the reaction changes as the reaction takes place.

f) SUPPLEMENT Use the collision theory to explain the change in the rate of reaction.

g) The reaction was repeated using the same volume and concentration of hydrochloric acid and with the same mass of marble, but as a powder instead of chips. Draw another curve on your graph paper, using the same axes as before (label it as Graph 2), to show how the original results will change.

h) The reaction was repeated, but this time using the original mass of new marble chips and the same volume of hydrochloric acid, but with the acid only half as concentrated as originally. Draw another curve on your graph paper, using the same axes as before (label it as Graph 3), to show how the original results will change.

5. SUPPLEMENT For a chemical reaction to occur, the reacting particles must collide. Explain why not all collisions between the particles of the reactants lead to a chemical reaction.

6. SUPPLEMENT The diagrams below show the activation energies of two different reactions A and B.

a) Explain the *activation energy* of a reaction.

b) Select which reaction is likely to have the higher rate of reaction at a particular temperature. Explain your answer.

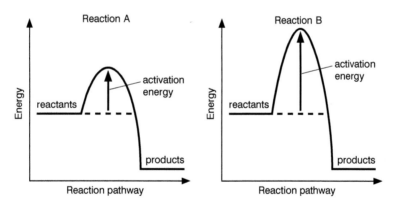

c) Explain how the presence of a catalyst in a reaction increases the rate of a reaction.

Redox reactions

△ Fig. 6.16 Oxidation and reduction are both taking place in this bonfire.

INTRODUCTION

Oxidation reactions are very familiar in everyday life – like the rusting of iron and bleaching, which is effective because bleach is a powerful oxidising agent. Whenever anything burns, an oxidation reaction takes place between the fuel and oxygen in the air. Reduction reactions may seem less familiar, but oxidation and reduction go hand in hand – if an element or compound in a chemical reaction is oxidised, then another element or compound in the same reaction must be reduced. So even when a bonfire is burning furiously and using oxygen from the air, reduction is taking place at the same time!

KNOWLEDGE CHECK

✓ Ions are charged atoms and have particular charges.
✓ Chemical equations and associated state symbols provide details about particular reactions.
✓ **SUPPLEMENT** Half-equations are used to explain the changes taking place in a reaction.

LEARNING OBJECTIVES

✓ Define redox reactions as involving simultaneous oxidation and reduction.
✓ Define oxidation as gain of oxygen and reduction as loss of oxygen.
✓ Identify redox reactions as reactions involving gain and loss of oxygen.
✓ Identify oxidation and reduction in redox reactions. (Oxidation number limited to its use to name ions, e.g. iron(II), iron(III), copper(II).)
✓ **SUPPLEMENT** Define oxidation in terms of: loss of electrons; an increase in oxidation number (determination of oxidation numbers is **not** required).
✓ **SUPPLEMENT** Define reduction in terms of: gain of electrons; a decrease in oxidation number (determination of oxidation numbers is **not** required).

OXIDATION, REDUCTION AND REDOX

When oxygen is added to an element or a compound, the process is called **oxidation**:

$2Cu(s) + O_2(g) \rightarrow 2CuO(s)$

The copper has been oxidised.

Removing oxygen from a compound is called **reduction**:

$CuO(s) + Zn(s) \rightarrow ZnO(s) + Cu(s)$

The copper(II) oxide has been reduced.

If we look more carefully at this last reaction, we see the zinc has changed to zinc oxide: that is, it has been oxidised at the same time as the copper(II) oxide has been reduced.

This is one example of reduction and oxidation taking place at the same time, in the same reaction. Reactions like these that involve the simultaneous gain and loss of oxygen are called **redox** reactions.

SUPPLEMENT

There is another way to look at redox reactions if we consider the reaction in a different way:

$$CuO(s) \quad + \quad Zn(s) \quad \rightarrow \quad ZnO(s) \quad + \quad Cu(s)$$

ionic compound   element   ionic compound   element

Rewrite it:

$$Cu^{2+} + O^{2-} + Zn \rightarrow Zn^{2+} + O^{2-} + Cu$$

Remove the oxygen ions because they are on both sides of the equation (the O^{2-} ion is unchanged):

$$Cu^{2+}(s) + Zn(s) \rightarrow Zn^{2+}(s) + Cu(s)$$

Split the equation into two half-equations and add electrons to balance them:

$$Cu^{2+} + 2e^- \rightarrow Cu$$

that is: CuO to Cu = reduction

$$Zn \rightarrow Zn^{2+} + 2e^-$$

that is: Zn to ZnO = oxidation

In this reaction the copper(II) oxide has been reduced by the zinc. The zinc is a reducing agent.

The zinc itself has been oxidised by the copper(II) oxide. The copper(II) oxide is an oxidising agent and is itself reduced.

You now have a new definition, using electrons instead of oxygen:

- Oxidation is loss of electrons.
- Reduction is gain of electrons.

Remember this as OIL-RIG: Oxidation *Is* Loss – Reduction *Is* Gain.

Oxidation numbers

When you learned to write chemical formulas, you were introduced to the use of Roman numerals for metals that had more than one ion – for example, iron as Fe^{2+} or Fe^{3+}:

- iron(II) oxide = FeO
- iron(III) oxide = Fe_2O_3

The II and III are called **oxidation numbers**.

- Fe^{2+} has an oxidation number of +2
- Fe^{3+} has an oxidation number of +3
- O^{2-} has an oxidation number of −2

You take the ion charge and reverse it, so an ion of 3− has oxidation number −3.

The oxidation number of elements is always 0 (zero).

An oxidation number describes how many electrons an atom loses or gains when it forms a chemical bond.

If you look at the equation for the reaction between Cu^{2+} and Zn again, you will see that we can use oxidation numbers:

$Cu^{2+} + 2e^- \rightarrow Cu$ reduction

+2   0

$Zn \rightarrow Zn^{2+} + 2e^-$ oxidation

0   +2

This gives another definition for oxidation/reduction:

• Oxidation involves an increase in oxidation number.
• Reduction involves a decrease in oxidation number.

SCIENCE IN CONTEXT

PHOTOCHROMIC GLASS

Some people who wear glasses prefer those with photochromic lenses, which darken when exposed to bright light. These glasses eliminate the need for sunglasses – they can reduce up to 80% of the light transmitted through the lenses to the eyes. The basis of this change in colour in response to light can be explained in terms of redox reactions.

△ Fig. 6.17 A photochromic reaction was produced when the right lens of these glasses was exposed to bright light.

Glass is ordinarily transparent to visible light. In photochromic lenses, silver chloride (AgCl) and copper(I) chloride (CuCl) crystals are added during the manufacturing of the glass. These crystals become uniformly embedded in the glass.

One characteristic of silver chloride is that it is affected by light. The following reactions occur:

$Cl^- \rightarrow Cl + e^-$ = oxidation
$Ag^+ + e^- \rightarrow Ag$ = reduction

The chloride ions are oxidised to produce chlorine atoms and the silver ions are reduced to silver atoms. The silver atoms cluster together and block the transmission of light, causing the lenses to darken. This process occurs almost instantaneously.

The photochromic process would not be useful unless it were reversible. The presence of copper(I) chloride reverses the darkening process in the following way. When the lenses are removed from bright light, the following reaction occurs:

$Cl + Cu^+ \rightarrow Cu^{2+} + Cl^-$

The chlorine atoms formed by the exposure to light are reduced by the copper(I) ions forming chloride ions (Cl^-). The copper(I) ions are oxidised to copper(II) ions. The copper(II) ions then oxidise the silver atoms:

$$Cu^{2+} + Ag \rightarrow Cu^+ + Ag^+$$

The result of these reactions is that the lenses become transparent again as the silver atoms are converted back to silver ions.

Challenge Question: Which substances are oxidised and reduced in the equation:

$$Cu^{2+} + Ag \rightarrow Cu^+ + Ag^+$$

QUESTIONS

1. Explain the term *reduction*.

2. Write a word equation to show the oxidation of aluminum when it reacts with oxygen in the air.

3. SUPPLEMENT In the following word equation identify what has been oxidised and what has been reduced.

   Copper(II) oxide + carbon → copper + carbon dioxide

4. SUPPLEMENT In the following half-equation, has the chromium (Cr) atom been oxidised or reduced? Explain your answer.

   $Cr(s) \rightarrow Cr^{3+}(aq) + 3e^-$

Key terms

oxidation, oxidation number, redox, reduction

During your study of this topic you should have learned:

○ How to describe the definitions of oxidation and reduction in terms of oxygen loss or gain.

○ That oxidation numbers are used to name transition metals, ions, for example, iron(II), iron(III), copper(II).

○ That redox reactions involve simultaneous reduction and oxidation.

○ How to identify oxidation and reduction in redox reactions.

○ **SUPPLEMENT** The definition of redox in terms of the gain or loss of electrons.

○ **SUPPLEMENT** How to distinguish between oxidation and reaction reactions from the change in oxidation number.

End of topic questions

1. Select which of the following reactions shows a reduction reaction.

A $2CO + O_2 \rightarrow 2CO_2$

B $CO_2 + C \rightarrow 2CO$

C $NaOH + HCl \rightarrow NaCl + H_2O$

D $CuSO_4{\bullet}5H_2O \rightarrow CuSO_4 + 5H_2O$

2. The following equation shows a redox reaction:

$Mg(s) + ZnO(s) \rightarrow MgO(s) + Zn(s)$

a) State what has been oxidised in the reaction.

b) State what has been reduced in the reaction.

c) Explain why this reaction is a redox reaction.

3. **SUPPLEMENT** The relationship between lead atoms and lead ions is shown in the following half-equation:

$Pb(s) \rightarrow Pb^{2+}(s) + 2e^-$

In this reaction, state whether the lead atom has been oxidised or reduced. Explain your answer.

4. **SUPPLEMENT** The reaction between copper and chlorine produces copper(II) chloride.

$Cu(s) + Cl_2(g) \rightarrow CuCl_2(s)$

Is this reaction a redox reaction? Explain your answer.

In this section you will explore the reactions of acids, bases and alkalis (which are bases that dissolve in water). In addition, you will learn about salts, which are made from the reaction of acids with bases and alkalis. You will probably have studied acids and alkalis previously and will certainly be familiar with their identification using indicators. A study of salts may be new to you, although it is very likely you have encountered common salt as sodium chloride. Once again you will be using chemical equations, as studied in Section 3, Stoichiometry.

Starting points

1. Give the names of the common acids you are familiar with.

2. Do you know the names of some common alkalis?

3. What indicators have you encountered in your previous work?

4. Do you know what name is given to the reaction of an acid with an alkali?

SYLLABUS SECTIONS COVERED

7.1 The characteristic properties of acids and bases

7.2 Oxides

7.3 Preparation of salts

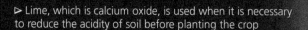

▷ Lime, which is calcium oxide, is used when it is necessary to reduce the acidity of soil before planting the crop

7

Acids, bases and salts

△ Fig. 7.1 Sodium hydroxide.

Acids, bases and salts

INTRODUCTION

Acids are commonly used in everyday life. Many of them, such as hydrochloric acid and sulfuric acid, are extremely toxic and corrosive. About 20 million tonnes of hydrochloric acid are manufactured worldwide each year. Some of this is used to make important chemicals such as PVC (polyvinyl chloride) plastic. Alkalis and bases are less common in everyday use, yet about 60 million tonnes of sodium hydroxide are produced worldwide each year and used in the manufacture of paper and soap. Sodium hydroxide is harmful and corrosive. Common salt, sodium chloride, is an example of a salt.

KNOWLEDGE CHECK

✓ Common acids include hydrochloric acid and sulfuric acid.
✓ Vegetable dyes can be used as indicators to identify acids and alkalis.
✓ State symbols include (s), (l), (g) and (aq).

LEARNING OBJECTIVES

✓ Describe the characteristic properties of acids in terms of their reactions with metals, bases and carbonates.
✓ Describe acids in terms of their effect on litmus and methyl orange.
✓ State that bases are oxides or hydroxides of metals and that alkalis are soluble bases.
✓ Describe the characteristic properties of bases in terms of their reactions with acids.
✓ Describe alkalis in terms of their effect on litmus and methyl orange.
✓ Describe how to compare, neutrality, relative acidity and relative alkalinity in terms of colour and pH (using universal indicator paper).
✓ Describe the neutralisation reaction between an acid and an alkali to produce a salt and water. (The ionic equation for this reaction is **not** required.)
✓ Classify oxides as acidic, including SO_2 and CO_2, or basic, including CuO and CaO, related to metallic and non-metallic character.
✓ SUPPLEMENT Describe amphoteric oxides as oxides that react with acids and with bases to produce a salt and water.
✓ SUPPLEMENT Classify Al_2O_3 and ZnO as amphoteric oxides.
✓ Describe the preparation, separation and purification of soluble salts by reaction of an acid with: an alkali by titration; excess metal; excess insoluble base; excess insoluble carbonate (students do not need to know general solubility rules for salts).
✓ Define a hydrated substance as a substance that is chemically combined with water and an anhydrous substance as a substance containing no water.
✓ SUPPLEMENT Describe the preparation of insoluble salts by precipitation (students do not need to know general solubility rules for salts).

THE CHARACTERISTIC PROPERTIES OF ACIDS AND BASES

Aqueous solutions

When any substance dissolves in water, it forms an aqueous solution, shown by the state symbol (aq). Aqueous solutions can be acidic, alkaline or neutral. A neutral solution is neither acidic nor alkaline.

Indicators are used to tell if a solution is acidic, alkaline or neutral. They can be used either as liquids or in paper form, and they turn different colours with different solutions. There are different indicators that can be used.

The most common indicator is **litmus**. Its colours are shown in Table 7.1.

Colour of litmus	Type of solution
Red	Acidic
Blue	Alkaline

Δ Table 7.1 Litmus.

Universal indicator can show how strongly acidic or how strongly alkaline a solution is because it has more colours than litmus. Each colour is linked to a number ranging from 0 (most strongly acidic solution) to 14 (most strongly alkaline solution). A neutral substance has a pH of 7. This range is called the **pH scale**.

pH	Examples of solutions and their respective pH
14	Liquid drain cleaner, caustic soda
13	Bleaches, oven cleaner
12	Soapy water
11	Household ammonia (11.9)
10	Milk of magnesia (10.5)
9	Toothpaste (9.9)
8	Baking soda (8.4), seawater, eggs
7	Pure water (7)
6	Urine (6), milk (6.6)
5	Acid rain (5.6), black coffee (5)
4	Tomato juice (4.1)
3	Grapefruit and orange juice, soft drinks
2	Lemon juice (2.3), vinegar (2.9)
1	Hydrochloric acid secreted from the stomach lining (1)
0	Battery acid

Δ Fig. 7.2 The pH scale.

Methyl orange indicator is red in acid solution and yellow in alkaline solution.

What are acids?

Acids are substances that contain replaceable hydrogen atoms. These hydrogen atoms are replaced in chemical reactions by metal atoms, forming a compound known as a salt. Acids have pHs in the range 0–6 (must be less than 7 as this corresponds to neutral).

Acid name	Acid formula
Hydrochloric acid	HCl
Nitric acid	HNO_3
Sulfuric acid	H_2SO_4
Phosphoric acid	H_3PO_4

△ Table 7.2 Common acids.

The typical reactions of acids include:

- Acid + metal makes a salt and hydrogen gas.
- Acid + carbonate makes a salt, carbon dioxide and water.
- Acid + alkali makes a salt and water.

What are bases and alkalis?

The oxides and hydroxides of metals are called **bases**.

If the oxide or hydroxide of a metal dissolves in water, it is also called an **alkali**. For example:

sodium + oxygen → sodium oxide

$$4Na(s) + O_2(g) \rightarrow 2Na_2O(s)$$

sodium oxide + water → sodium hydroxide

$$Na_2O(s) + H_2O(l) \rightarrow 2NaOH(aq)$$

Sodium oxide is a base because it is the oxide of the metal sodium. In addition, it reacts with water to make the alkali sodium hydroxide. Alkalis have pHs in the range 8–14 (must be more than 7 as this corresponds to neutral).

The typical reactions of bases include:

- Base + acids make a salt and water.

QUESTIONS

1. Two solutions of the same concentration are tested with universal indicator paper. Solution A has a pH of 8 and solution B has a pH of 14. Explain what this tells you about the two solutions.

2. Methyl orange is added to a solution and the solution turns red. Explain what this tells you about the solution.

3. Calcium oxide is an example of a base. Suggest how we know this.

THE IMPORTANCE OF CONTROLLING ACIDITY IN SOIL

Different plants grow better in different types of soil. The pH of a soil is an important factor in the growth of different plants – some plants prefer slightly acidic conditions and others slightly alkaline conditions.

If soil is too acidic then it can be neutralised using quicklime. Quicklime is made from limestone, which is quarried from limestone rocks. It is heated in lime kilns at 1200 °C to make calcium oxide or quicklime:

limestone $\xrightarrow{1200\,°C}$ quicklime + carbon dioxide

$CaCO_3(s) \rightarrow \quad CaO(s) \quad + \quad CO_2(g)$

When quicklime is added to water, it makes calcium hydroxide, which is an alkali and so can neutralise the acidic soil:

quicklime + water $\rightarrow$ slaked lime

$CaO(s) \quad + \quad H_2O(l) \rightarrow Ca(OH)_2(s)$

Adding fertilisers to soil can also affect the pH and so the soil may have to be treated by adding acids or alkalis. The pH of soil can be measured by taking a small sample, putting it in a test-tube with distilled water and adding indicator solution or using indicator paper. The pH can be found from a pH chart.

Challenge Question: Calculate how much quicklime (calcium oxide) can be made from 1 kg of limestone (calcium carbonate). (A_r: C = 12, O = 16, Ca = 40)

OXIDES

The **oxides** of elements can often be made by heating the element in air or oxygen. For example, the metal magnesium burns in oxygen to form magnesium oxide:

magnesium + oxygen $\rightarrow$ magnesium oxide

$2Mg(s) \quad + O_2(g) \quad \rightarrow 2MgO(s)$

Magnesium oxide forms as a white solid. When distilled water is added to the ash and the mixture is tested with universal indicator, the pH is greater than 7 – the oxide has formed an alkaline solution.

△ Fig. 7.3 Magnesium burning in oxygen.

△ Fig. 7.4 Magnesium oxide is a white solid.

When sulfur is burned in oxygen, sulfur dioxide gas is formed:

sulfur + oxygen → sulfur dioxide

$$S(s) + O_2(g) \rightarrow SO_2(g)$$

When sulfur dioxide is dissolved in water and then tested with universal indicator solution, the pH is less than 7 – the oxide has formed an acidic solution.

The oxides of most elements can be classified as either **basic oxides** or **acidic oxides**. Copper(II) oxide and calcium oxide are examples of basic oxides. Sulfur dioxide and carbon dioxide are examples of acidic oxides. Some elements form neutral oxides. For example, water is a neutral oxide. Basic oxides that dissolve in water are called alkalis.

Oxides that do not dissolve in water cannot be identified using the pH of their solutions. For insoluble oxides, the test is seeing if they will react with hydrochloric acid, HCl (these would be basic oxides) or sodium hydroxide solution, NaOH (these would be acidic oxides). Oxides that do not react with hydrochloric acid or sodium hydroxide are neutral oxides.

Most metal oxides are basic oxides, such as CaO, MgO, CuO, Na_2O. Basic oxides react with acids (neutralisation):

acid + base → a salt + water

$$2HCl(aq) + MgO(s) \rightarrow MgCl_2(aq) + H_2O(l)$$

Most non-metal oxides are acidic, for example:

NO_2, SO_2, SO_3, CO_2, P_2O_5

Acidic oxides dissolve in water to form acids, for example sulfuric acid (H_2SO_4):

$$SO_3(g) + H_2O(l) \rightarrow H_2SO_4(aq)$$

Oxide	Type of oxide	pH of solution	Other reactions of the oxide
Metal oxide	Basic	More than 7 (alkaline)	Reacts with an acid to make a salt + water
Non-metal oxide	Acidic	Less than 7 (acidic)	Reacts with a base to make a salt + water

△ Table 7.3 Acidic and basic oxides.

SUPPLEMENT

Amphoteric oxides are the oxides of less reactive metals, for example, Al_2O_3 and ZnO. They can behave as both acidic oxides *and* basic oxides, so react with both bases and acids.

QUESTIONS

1. Suggest whether potassium oxide is a basic oxide or acidic oxide. Explain your answer.

2. SUPPLEMENT Contrast the difference between a basic oxide and an amphoteric oxide.

3. Balance the following equation for the reaction of potassium with oxygen to form potassium oxide:
$$\_K(s) + \_O_2(g) \rightarrow \_K_2O(s)$$

ACIDS, BASES AND SALTS

SALTS

Acids contain replaceable hydrogen atoms. When metal atoms take the place of hydrogen atoms, a compound called a **salt** is formed. The names of salts have two parts, as shown in Fig. 7.5.

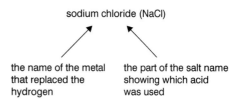

sodium chloride (NaCl)

the name of the metal that replaced the hydrogen

the part of the salt name showing which acid was used

△ Fig. 7.5 A salt – sodium chloride.

Table 7.4 shows the four most common acids and their salt names.

Acid	Salt name (ion)
Hydrochloric (HCl)	Chloride (Cl^-)
Nitric (HNO_3)	Nitrate (NO_3^-)
Sulfuric (H_2SO_4)	Sulfate (SO_4^{2-})
Phosphoric (H_3PO_4)	Phosphate (PO_4^{3-})

△ Table 7.4 Common acids and their salt names.

△ Fig. 7.6. Sodium chloride crystals.

△ Fig. 7.7 Copper(II) sulfate crystals.

Salts are ionic compounds. The names of these compounds are created by taking the first part of the name from the metal ion, which is a positive ion (cation), and the second part of the name from the acid, which is a negative ion (anion). For example:

copper(II) sulfate: Cu^{2+} and $SO_4^{2-} \rightarrow CuSO_4$

cation   anion   salt

Making salts

There are five common methods for making salts. Four of these make soluble salts and one makes insoluble salts.

REMEMBER

Some salts are chemically combined with water – they are **hydrated**. A salt that is not combined with water is **anhydrous**. A good example of this is copper(II) sulfate. As $CuSO_4$ the salt is anhydrous. As $CuSO_4 \bullet 5H_2O$ the salt is hydrated.

Making soluble salts

1.

acid + alkali → a salt + water

For example:

$HCl(aq)$ + $NaOH(aq)$ → $NaCl(aq)$ + $H_2O(l)$

2.

acid + base → a salt + water

For example:

$H_2SO_4(aq)$ + $CuO(s)$ → $CuSO_4(aq)$ + $H_2O(l)$

3.

acid + carbonate → a salt + water + carbon dioxide

For example:

$2HNO_3(aq)$ + $CuCO_3(s)$ → $Cu(NO_3)_2(aq)$ + $H_2O(l)$ + $CO_2(g)$

4.

acid + metal → a salt + hydrogen

For example:

$2HCl(aq)$ + $Mg(s)$ → $MgCl_2(aq)$ + $H_2(g)$

Here is a shortcut for remembering the four general equations above. Remember the initials of the reactants:

A (acid) + A (alkali)

A (acid) + B (base)

A (acid) + C (carbonate)

A (acid) + M (metal)

The symbol '(aq)' after the formula of the salt shows that it is a soluble salt.

When exactly the right volumes of acid react with an alkali or base to produce a neutral solution, this is called neutralisation. An indicator is used to find out when **neutralisation** occurs.

QUESTIONS

1. Complete the word equations for the following reactions to form a salt:

    a) magnesium + sulfuric acid →

    b) copper(II) oxide + hydrochloric acid →

    c) copper(II) carbonate + nitric acid →

2. SUPPLEMENT Complete the following ionic equations:

    a) $Mg + 2H^+$ →

    b) $CO_3^{2-} + 2H^+$ →

In the laboratory

Of the four methods for making soluble salts, shown by the symbol (aq), only one uses two solutions:

1. acid(aq) + alkali(aq) → a salt(aq) + water(l)

The other three methods involve adding a solid(s) to a solution(aq):

2. acid(aq) + base(s) → a salt(aq) + water(l)

3. acid(aq) + carbonate(s) →
a salt(aq) + water(l) + carbon dioxide(g)

4. acid(aq) + metal(s) →
a salt(aq) + hydrogen(g)

Method 1 involves the **titration** method to accurately measure volumes of the acid and alkali solutions. An indicator is used to show when exact quantities of acid and alkali have been mixed. The procedure is then repeated using the same exact volumes of acid and alkali, but without the indicator. The resulting solution is evaporated to the point of **crystallisation**, then left to cool and the salt to crystallise.

△ Fig. 7.8 Using the neutralisation method for a titration.

The general procedure used for each of the methods 2, 3 and 4 is the same:

- The solid (insoluble base, insoluble carbonate or metal) is added to the acid with stirring until no more solid will react. The solid is in excess. Heating may be necessary.
- The mixture is filtered to remove unreacted solid and the solution is collected as the **filtrate** in an evaporating dish.
- The solution is evaporated to the point of crystallisation and is then left to cool and the salt to crystallise.

The process is summarised in Fig. 7.9:

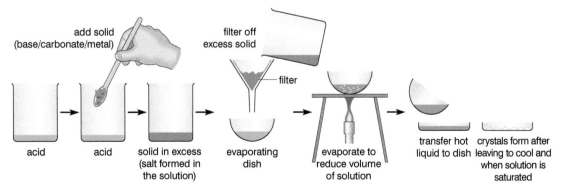

△ Fig. 7.9 Making soluble salts from solids.

1. Explain what a salt is.

2. Which acid would you use to make a sample of sodium sulfate?

3. What is the name of the salt formed when calcium carbonate reacts with nitric acid?

4. Explain the term *neutralisation*.

SUPPLEMENT

Making insoluble salts

If two solutions of **soluble** salts are mixed together to form two new salts, and one of the products is **insoluble**, the insoluble salt forms a precipitate – a solid made in solution. This process is called **precipitation**. The general equation is:

soluble salt + soluble salt → insoluble salt (precipitate) + soluble salt

For example:

$$Na_2CO_3(aq) + CuSO_4(aq) \rightarrow CuCO_3(s) + Na_2SO_4(aq)$$

The state symbols show the salts in solution as (aq) and the precipitate – the insoluble salt – as (s).

In the laboratory

The practical method involves making a precipitate of an insoluble salt by mixing solutions of two soluble salts.

The procedure is as follows:

1. The two solutions of soluble salts are mixed together and a precipitate forms.

2. The precipitate is separated by filtration.

3. The precipitate is then washed with a little cold water and allowed to dry.

The process is summarised in Fig. 7.10.

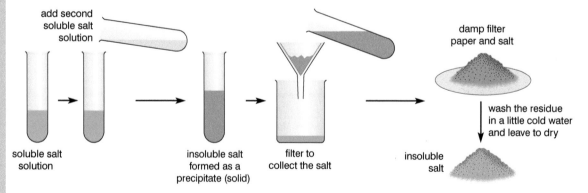

add second
soluble salt
solution

damp filter
paper and salt

soluble salt
solution

insoluble salt
formed as a
precipitate (solid)

filter to
collect the salt

insoluble
salt

wash the residue
in a little cold water
and leave to dry

△ Fig. 7.10 Making an insoluble salt.

QUESTIONS

1. **SUPPLEMENT** Explain the term *precipitation* reaction.

2. **SUPPLEMENT** When preparing an insoluble salt state what process is used to separate the insoluble salt from any soluble salts.

3. **SUPPLEMENT** Explain why the insoluble salt is washed with a little cold water before it is left to dry.

4. **SUPPLEMENT** Lead(II) chloride is an insoluble salt and can be prepared by mixing solutions of lead(II) nitrate and sodium chloride.

   a) Write a word equation for the reaction.

   b) Write a balanced equation with state symbols for the reaction.

SCIENCE IN CONTEXT

SOME INTERESTING FACTS ABOUT ACIDS AND ALKALIS

1. Pure sulfuric acid is a clear, oily, highly corrosive liquid. It was well known to Islamic, Greek and Roman scholars in the ancient world, when it was called 'oil of vitriol'. Although pure sulfuric acid does not occur naturally on Earth because of its attraction for water, dilute sulfuric acid is found in acid rain and in the upper atmosphere of the planet Venus. It has a wide range of industrial uses from making fertilisers, dyes, paper and pharmaceuticals to batteries, steel and iron. Sulfuric acid

△ Fig. 7.11 A wasp.

is not toxic but it is highly reactive with water, in a strongly exothermic reaction. It can cause severe burns. The acid must be stored in glass containers (never plastic or metal) and handled with extreme care.

2. Hydrochloric acid, although classified as toxic and corrosive, is part of the gastric acid in the stomach and is involved in digestion. Excess acid in the stomach can cause indigestion but 'anti-acid' (alkali) medications can be taken to neutralise this.

3. Perhaps the strongest acid is a mixture of nitric acid and hydrochloric acid, known as 'aqua regia' because it reacts with the 'royal' metals. Unlike other acids, it reacts with very unreactive metals such as gold and platinum. However, some metals like titanium and silver are not affected.

4. Formic acid (now called methanoic acid) is in the venom of ant and bee stings. Such stings can be relieved by (or neutralised) with an alkali such as sodium bicarbonate (sodium hydrogencarbonate). Wasp stings, however, contain an alkali and so need to be neutralised by a weak acid such as vinegar.

The differences can be remembered using:

Bee – Bicarb

Wasp (W looks like two Vs) – Vinegar

Challenge Question: You can buy liquids which can be used to soothe skin after insect bites and stings. One particular product contains the following ingredients: ammonia, liquid paraffin and water. Consider whether you would use this product to soothe your skin after a wasp sting. Explain your answer.

Developing practical skills

A student wanted to make a sample of copper(II) sulfate crystals, $CuSO_4 \bullet 5H_2O$. She used the following steps in her method.

She put on eye protection and warmed 50 cm³ of dilute sulfuric acid in a 250 cm³ beaker and then copper(II) oxide was added a spatula at a time, stirring the reaction mixture with a glass rod.

When no more copper(II) oxide would react, she filtered the mixture and collected the copper(II) sulfate solution in an evaporating basin.

She then heated the solution in the evaporating basin until she could see crystals starting to form and then allowed it to cool.

After several hours, blue crystals of copper(II) sulfate had formed. She drained off any remaining liquid and dried the crystals between filter papers. She washed her hands after she had finished.

Using and organising techniques, apparatus and materials

1. Copper(II) sulfate crystals are 'hydrated'. Explain what this means.

2. Draw a diagram showing the apparatus the student could have used to filter the mixture.

3. Give the general name given to a liquid that passes through a filter paper.

4. Explain how the student could have tested for the crystallisation point while heating the filtrate.

Observing, measuring and recording

5. State the colour of copper(II) oxide.

6. Describe the colour of the solution the student evaporated.

Evaluating methods

7. While heating the solution in the evaporating dish the student noticed some very pale blue powder around the edges of the evaporating basin. Suggest what you think this powder was and explain how it had formed.

End of topic checklist

Key terms

acid, acidic oxide, alkali, anhydrous, base, basic oxide, crystallisation, filtrate, hydrated, indicator, insoluble, litmus, neutralisation, oxide, pH scale, salt, soluble, titration, universal indicator

SUPPLEMENT amphoteric oxide, precipitation

During your study of this topic you should have learned:

○ How to describe the characteristic properties of acids as reactions with metals, bases, carbonates and the effects on litmus and methyl orange.

○ How to describe the characteristic properties of bases (oxides or hydroxides of metals) as reactions with acids and with and the effects on litmus and methyl orange.

○ How to compare neutrality, relative acidity and alkalinity using universal indicator.

○ How to classify oxides as either acidic (SO_2 and CO_2) or basic (CuO or CaO), related to metallic and non-metallic character.

○ How to describe the preparation, separation and purification of soluble salts as prepared by the following reactions:
   ● acid + alkali (titration)
   ● acid + metal
   ● acid + insoluble base
   ● acid + insoluble carbonate.

○ How to define the terms hydrated and anhydrous.

○ **SUPPLEMENT** How to define amphoteric oxides, with Al_2O_3 and ZnO as examples.

○ **SUPPLEMENT** How to describe the preparation of insoluble salts by precipitation.

End of topic questions

1. Which one of these statements about hydrochloric acid is correct?

A It has a pH of 7.

B It turns litmus paper blue.

C It causes methyl orange to turn yellow.

D It has a pH of 1.

2. **a)** Explain what an indicator is.

b) State what the pH scale is.

c) State what the following pH numbers indicate about a solution that has been tested:

i) pH 6

ii) pH 8

iii) pH 14.

3. **a)** Define the term 'acid'.

b) Define the term 'alkali'.

c) State the name of the process when an acid reacts with an alkali to form water.

4. Calcium chloride, a soluble salt, can be made from calcium oxide and dilute hydrochloric acid.

a) State what type of chemical calcium oxide is.

b) State what type of chemical calcium chloride is.

c) Describe the different stages in the preparation of calcium chloride crystals.

d) **SUPPLEMENT** Write a fully balanced equation, including symbols, for the reaction between calcium oxide and dilute hydrochloric acid.

5. **SUPPLEMENT** Barium sulfate is an insoluble salt and can be made using a precipitation reaction.

a) State what acid can be used to make barium sulfate.

b) State what other chemical could be used to make barium sulfate.

c) Describe the different stages in the preparation of a dry sample of barium sulfate.

d) Write a fully balanced equation, including state symbols, for the reaction to make barium sulfate.

6. **SUPPLEMENT** Copy and complete the following equations and include state symbols:

a) $2KOH(aq) + H_2SO_4(aq) \rightarrow$ —— + ——

b) $2HCl(aq) + MgO(s) \rightarrow$ —— + ——

c) $2HNO_3(aq) + BaCO_3(s) \rightarrow$ —— + ——

d) $2HCl(aq) + Zn(s) \rightarrow$ —— + ——

e) $ZnCl_2(aq) + K_2CO_3(aq) \rightarrow$ —— + ——

Practice questions for Sections 4, 5, 6 and 7

Example answer

Note: Practice questions, sample answers and comments have been written by the authors. The marks awarded for these questions indicate the level of detail required in the answers. In examinations, the way marks are awarded may be different. References to assessment and/or assessment preparation are the publisher's interpretation of the syllabus requirements and may not fully reflect the approach of Cambridge Assessment International Education.

Question 1 SUPPLEMENT

Solutions of lead(II) nitrate and potassium iodide react together to make the insoluble substance lead(II) iodide.

The equation for the reaction is

$$Pb(NO_3)_2(aq) + 2KI(aq) \rightarrow 2KNO_3(aq) + PbI_2(s)$$

An investigation was carried out to find how much precipitate formed with different volumes of lead(II) nitrate solution.

A student measured out 15 cm³ of potassium iodide solution using a measuring cylinder.

He poured this solution in to a clean boiling tube.

Using a clean measuring cylinder, he measured out 2 cm³ of lead(II) nitrate solution (of the same concentration as the potassium iodide solution). He added this to the potassium iodide solution.

A cloudy yellow mixture formed and the precipitate was left to settle.

The student then measured the height (in cm) of the precipitate using a ruler.

The student repeated the experiment using different volumes of lead(II) nitrate solution. The graph shows the results obtained.

COMMENTS

a) **i)** Correct point marked.

ii) Correct explanation. Also correct would be tube not being vertical when being set up so precipitate not level.

iii) Correct response.

b) **i)** Correct response.

ii) Correct response – also correct is 'lead(II) nitrate in excess'.

c) **i)** Correct reading of ruler.

ii) Answer is 3.9 cm³ – the horizontal axis scale has been misread and so the incorrect unit has been given.

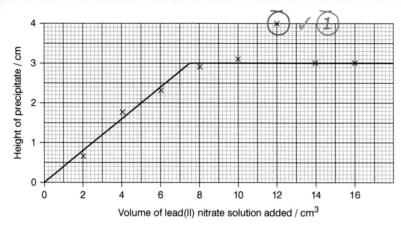

Volume of lead(II) nitrate solution added / cm³

a) i) On the graph, circle the point that seems to be anomalous. (1)

ii) Explain two things that the student may have done in the experiment to give this anomalous result.

Precipitate not settled ✓ ① *Because not left long enough* ✓ ① (2)

iii) Explain why the graph line must go through (0, 0).

Cannot have a precipitate if no lead nitrate added yet. ✓ ① (1)

b) Suggest a reason why the height of the precipitate stops increasing.

No more potassium iodide left to react. ✓ ① (1)

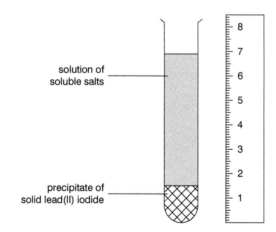

solution of soluble salts

precipitate of solid lead(II) iodide

c) i) How much precipitate has been made in the tube drawn on the right?

1.5 cm ✓ ① (1)

ii) Use the graph to find the volume of lead(II) nitrate solution needed to make this amount of precipitate.

2.9 cm ✗ (1)

d) After he had plotted the graph, the student decided he should obtain some more results.

   i) Suggest what volumes of lead(II) nitrate solution he should use.

     Between 6 cm³ and 10 cm³ ✓ ①   (1)

   ii) Explain why he should use these volumes.

     Need to know exactly where the graph levels off ✓ ①   (1)

e) Suggest a different method for measuring the amount of precipitate formed.

   Filter ✓ ① *off each precipitate and weigh it* ✓ ①   (4)

(Total 13 marks)

d) i) Correct response.

   ii) Correct – this is the purpose of the experiment.

e) The answer could have been improved by noting that the filtered-off precipitate needs to be washed and 'dried' before being weighed.

Question 2

Copper(II) oxide reacts with carbon to form copper and carbon dioxide.

$$2CuO(s) + C(s) \rightarrow 2Cu(s) + CO_2(g)$$

Identify which of the following statements is correct.

A The carbon has been reduced.

B The copper(II) oxide has been oxidised.

C The copper(II) oxide has been reduced.

D The carbon dioxide has been oxidised.   (1)

(Total 1 mark)

Question 3 SUPPLEMENT

Dilute nitric acid reacts with marble chips to produce carbon dioxide. The equation is given below:

$$2HNO_3(aq) + CaCO_3(s) \rightarrow Ca(NO_3)_2(aq) + H_2O(l) + CO_2(g)$$

Some students investigated the effect of changing the temperature of the nitric acid on the rate of the reaction. The method is:

- Use a measuring cylinder to pour 50 cm³ of dilute nitric acid into a conical flask.
- Heat the acid to the required temperature.
- Put the flask on the balance.

- Add 15 g (an excess) of marble chips to the flask.
- Time how long it takes for the mass to decrease by 1.00 g.
- Repeat the experiment at different temperatures.

The students' results are shown in the table.

Temperature of acid (°C)	Time to lose 1.00 g (s)
20	93
33	68
44	66
55	40
67	30
76	25

a) i) Draw a graph of the results. (3)

   ii) One of the points is inaccurate. Circle this point on your graph. (1)

   iii) Suggest a possible cause for this inaccurate result. (1)

b) Use the graph to determine the times taken to lose 1.00 g at 40 °C and 60 °C. (2)

c) The rate of the reaction can be found using the equation:

$$\text{rate of reaction} = \frac{\text{mass lost}}{\text{time taken to lose mass}}$$

   i) Use this equation and your results from **b)** to calculate the rates of reaction at 40 °C and 60 °C. (2)

   ii) State the unit for these rates. (1)

   iii) State how the rate of reaction changes when the temperature increases. (1)

   iv) Explain in terms of particles and collisions why the rate changes when the temperature increases. (1)

(Total 12 marks)

Question 4 SUPPLEMENT

The diagram shows the apparatus used to electrolyse lead(II) bromide.

a) The wires connected to the electrodes are made of copper.

   Explain why copper conducts electricity. (1)

b) Explain why electrolysis does not occur unless the lead(II) bromide is molten. (2)

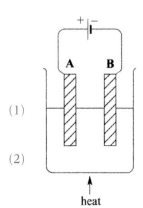

c) The reactions occurring at the electrodes can be represented by the equations shown in the table.

Copy and complete the table to show the electrode (A or B) at which each reaction occurs, and the type of reaction occurring (oxidation or reduction). (2)

Electrode reaction	Electrode	Type of reaction
$Pb^{2+} + 2e^- \rightarrow Pb$		
$2Br^- \rightarrow Br_2 + 2e^-$		

(Total 5 marks)

Question 5

Read the following instructions for the preparation of hydrated nickel(II) sulfate ($NiSO_4 \bullet 7H_2O$), then answer the questions that follow.

- Put 25 cm³ of dilute sulfuric acid in a beaker.

- Heat the sulfuric acid until it is just boiling and then add a small amount of nickel(II) carbonate.

- When the nickel carbonate has dissolved, stop heating, then add a little more nickel carbonate. Continue in this way until nickel carbonate is in excess.

- Filter the hot mixture into a clean beaker.

- Make the hydrated nickel(II) sulfate crystals from the nickel(II) sulfate solution.

The equation for the reaction is

$$NiCO_3(s) + H_2SO_4(aq) \rightarrow NiSO_4(aq) + CO_2(g) + H_2O(l)$$

a) State what piece of apparatus you would use to measure out 25 cm³ of sulfuric acid. (1)

b) Explain why the nickel(II) carbonate is added in excess. (1)

c) When nickel(II) carbonate is added to sulfuric acid, there is fizzing. Explain why. (1)

d) After filtration, which one of the following describes the nickel(II) sulfate in the beaker?

Select the correct answer.

crystals   filtrate   precipitate   water (1)

e) Explain how you would obtain pure dry crystals of hydrated nickel(II) sulfate from the solution of nickel(II) sulfate. (2)

(Total 6 marks)

In this section you will extend the work you did in Section 2, Atoms, elements and compounds, and focus in detail on the structure of the Periodic Table and the characteristic properties of particular groups of elements. If you look at the Periodic Table in the back of this book, you will see that it includes 118 elements. The good news is that you will not have to study the properties of all these elements! Because of the way elements have been arranged in the Periodic Table, learning about one element often provides a very good idea about how other elements may behave. You will study in some detail a group of metals and a group of non-metals, followed by elements known as 'transition elements' and a group known as the noble gases.

Starting points

1. What is an element – how would you define the term?

2. What does the proton number of an atom tell you about its structure?

3. In terms of electronic configuration what is the main difference between a metal and a non-metal?

4. Helium is a common noble gas. Do you know a use of helium?

SYLLABUS SECTIONS COVERED

3.1 Arrangement of elements

3.2 Group I properties

3.3 Group VII properties

3.4 Transition elements

3.5 Noble gases

▷ Many batteries contain lithium, which is a Group I metal.

8
The Periodic Table

Arrangement of elements

INTRODUCTION

With over 100 different **elements** in existence, it is very important to have some way of ordering them. The Periodic Table puts elements with similar properties into columns, with a gradual change in properties moving from left to right along the rows. This topic looks at some of the basic features of the Periodic Table. Later topics will look in more detail at particular groups and arrangements of the elements.

Δ Fig. 8.1 This ordering of elements was first published in 1871 by the Russian chemist Dmitri Mendeleev.

KNOWLEDGE CHECK

✓ All matter is made up of elements.
✓ The proton number of an element gives the number of protons (and the number of electrons) in an atom of the element.
✓ Electrons are arranged in shells around the nucleus of the atom.

LEARNING OBJECTIVES

✓ Describe the Periodic Table as an arrangement of elements in periods and groups and in order of proton number / atomic number.
✓ Describe the change from metallic to non-metallic character across a period.
✓ Explain similarities in the chemical properties of elements in the same group of the Periodic Table in terms of their electronic configuration.
✓ SUPPLEMENT Identify trends in groups, given information about the elements.

ARRANGEMENT OF ELEMENTS IN THE PERIODIC TABLE

As new elements were discovered in the 19th century, chemists tried to organise them into patterns based on the similarities in their properties. John Newlands classified elements according to their properties and Dmitri Mendeleev produced the classification system which is considered to be the basis of the modern Periodic Table. When the structure of the atom was better known, elements were arranged in order of increasing proton number, and then the patterns started to make more sense. (Proton/ atomic number is the number of protons in an atom.)

How are elements classified in the modern Periodic Table?

The modern **Periodic Table** arranges the elements in order of increasing proton number. They are then arranged in periods and groups so that elements with similar properties and reactions are shown close together. The complete Periodic Table on page 318 lists all the known elements. The simplified Periodic Table in Fig 8.2 shows the first 86 elements and the main groups and their names. The number above each element is the atomic number (proton number).

Groups I	II											III	IV	V	VI	VII	VIII
Periods																	
1						1 H hydrogen											2 He helium
2	3 Li lithium	4 Be beryllium										5 B boron	6 C carbon	7 N nitrogen	8 O oxygen	9 F fluorine	10 Ne neon
3	11 Na sodium	12 Mg magnesium				transition metals						13 Al aluminium	14 Si silicon	15 P phosphorus	16 S sulfur	17 Cl chlorine	18 Ar argon
4	19 K potassium	20 Ca calcium	21 Sc scandium	22 Ti titanium	23 V vanadium	24 Cr chromium	25 Mn manganese	26 Fe iron	27 Co cobalt	28 Ni nickel	29 Cu copper	30 Zn zinc	31 Ga gallium	32 Ge germanium	33 As arsenic	34 Se selenium	35 Br bromine
5	37 Rb rubidium	38 Sr strontium	39 Y yttrium	40 Zr zirconium	41 Nb niobium	42 Mo molybdenum	43 Tc technetium	44 Ru ruthenium	45 Rh rhodium	46 Pd palladium	47 Ag silver	48 Cd cadmium	49 In indium	50 Sn tin	51 Sb antimony	52 Te tellurium	53 I iodine
6	55 Cs caesium	56 Ba barium	57–71 lanthanoids	72 Hf hafnium	73 Ta tantalum	74 W tungsten	75 Re rhenium	76 Os osmium	77 Ir iridium	78 Pt platinum	79 Au gold	80 Hg mercury	81 Tl thallium	82 Pb lead	83 Bi bismuth	84 Po polonium	85 At astatine

(Group VIII / period entries: 36 Kr krypton, 54 Xe xenon, 86 Rn radon)

metal   non metal   transition metal   metalloid

Δ Fig. 8.2 The Periodic Table.

Periods

Horizontal rows of elements are arranged in increasing proton number from left to right. Rows correspond to **periods**, which are numbered from 1 to 7.

Moving across a period, each successive atom of the elements gains one proton and one electron (in the same outer shell).

You can see in Fig. 8.3 how the number of electrons in the outer shell increases across a period, for the elements in period 3.

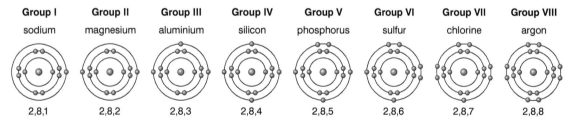

Group I	Group II	Group III	Group IV	Group V	Group VI	Group VII	Group VIII
sodium	magnesium	aluminium	silicon	phosphorus	sulfur	chlorine	argon
2,8,1	2,8,2	2,8,3	2,8,4	2,8,5	2,8,6	2,8,7	2,8,8

Δ Fig. 8.3 Moving across a period shows the electronic configuration of each element.

Moving across a period like Period 3 (sodium to argon), there is a change from metallic to non-metallic elements.

- Group I elements are the most reactive metal group, and as you go to the right the reactivity of the groups decreases. Group IV elements are the least reactive.

Groups

Vertical columns contain elements with the proton number increasing down the column – they are called **groups.** They are numbered from I to VIII (Group VIII is sometimes referred to as Group 0).

Groups are referred to as 'families' of elements because they have similar characteristics, just like families – the alkali metals (Group I), the alkaline earth metals (Group II) and the halogens (Group VII).

Elements in the same group do not have identical physical and chemical properties. For example, within a group there will be a trend (a gradual change between elements as you move down the group) in physical properties, such as melting point and density, and a trend in chemical reactivity. These group characteristics will be covered in more detail in the topics on Group 1 and Group VII elements.

REMEMBER

It is important to understand the relationship between group number, number of outer electrons, and metallic and non-metallic character across periods.

QUESTIONS

1. Find the element calcium in the Periodic Table. Answer these questions about calcium.

   a) Suggest its proton number.

   b) Describe what information the proton number gives about the structure of a calcium atom.

   c) State which group of the Periodic Table calcium is in.

   d) State which period of the Periodic Table calcium is in.

   e) Suggest whether calcium a metal or a non-metal.

2. What is the family name for the Group VII elements?

3. Suggest whether the Group VII elements are metals or non-metals.

Charges on ions and the Periodic Table

We can explain why elements in the same group have similar reactions in terms of the electron structures of their atoms. Elements with the same number of electrons in their outer shells have similar chemical properties.

Reactivities of elements

Going from the top to the bottom of a group in the Periodic Table, metals become more reactive, but non-metals become less reactive.

Group VIII elements, known as the noble gases, are very unreactive. They already have full outer electron shells and so rarely react with other elements to form compounds.

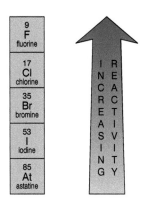

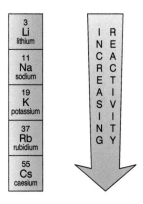

△ Fig. 8.4 The Group VII elements (non-metals) become more reactive further up the group.

△ Fig. 8.5 Group I elements (metals) become more reactive further down the group.

QUESTIONS

1. State how many electrons an aluminium atom has in its outer shell.

2. Would you expect elements in Group VI to be metals or non-metals? Explain your answer.

3. SUPPLEMENT State which is the most reactive element in Group VII.

SCIENCE IN CONTEXT

THE FIRST PERIODIC TABLE

In 1871 the Russian chemist Dmitri Mendeleev published his work on the Periodic Table. It included the 66 elements that were known at the time. Interestingly, Mendeleev left gaps in his arrangement when the next element in his order did not seem to fit. He predicted that there should be elements in the gaps but that they had yet to be discovered. One such element is gallium (discovered in 1875), which Mendeleev predicted would be between aluminium and indium.

In 2022 there were 118 known elements, but only 98 of these occurred naturally – the remaining 24 have been made artificially. Some elements are radioactive – for example the element americium (Am, proton number 95), which is used in smoke detectors.

Challenge Question: The element americium has an atomic number of 95 and an atomic mass of 243. Calculate how many protons, neutrons and electrons this atom has.

End of topic checklist

Key terms

Element, group, period, Periodic Table, proton number (atomic number)

During your study of this topic you should have learned:

○ How to describe the Periodic Table as a method of classifying elements into groups and periods and in order of increasing proton number/ atomic number.

○ How to describe the change from metallic to non-metallic character across a period.

○ That elements in the same group of the Periodic Table have similar properties due to their electronic configurations.

○ **SUPPLEMENT** How to identify trends in groups from information about the elements.

End of topic questions

1. Look at the diagram representing the simplified Periodic Table. The letters stand for elements.

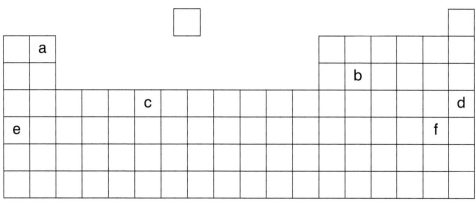

   a) State which element is in Group IV.

       A e   **B** c   **C** b   **D** d

   b) State which element is in the second period.

   c) State which element is a noble gas.

   d) State which element is a transition metal.

   e) State which elements are non-metals.

   f) State which element is most likely to be a gas.

2. Determine the electron configuration in the following atoms:

   a) sodium (proton number = 11)

   b) silicon (proton number = 14)

   c) fluorine (proton number = 9).

3. Describe how the metallic and non-metallic nature of the elements changes across Period 3 of the Periodic Table.

4. Explain why elements in the same group have similar chemical properties.

5. Compare and contrast the noble gases in Group VIII (or 0) with the other elements in Groups I to VII. Explain why the noble gases are so unreactive.

6. SUPPLEMENT The table provides information about the melting points and reactivity with cold water of three elements in Group 2.

Element	Melting point /°C	Reactivity with cold water
calcium	842	reacts to form bubbles of gas
strontium	777	reacts rapidly to form bubbles of gas
barium	727	reacts very vigorously producing bubbles of gas

   a) What trend is there in the melting points of the three Group 2 elements?

   b) What trend is there in the reactivity with cold water of the three Group 2 elements?'

Group I elements

△ Fig. 8.6 Potassium reacting with water.

INTRODUCTION

Metals are positioned on the left-hand side and in the middle of the Periodic Table. Therefore the Group I elements are metals, but they are rather different from the metals in everyday use. In fact, when you see how the Group I metals react with air and water, it is hard to think how they could be used outside the laboratory. This very high reactivity makes them interesting to study. Our focus is on the first three elements in the group: lithium, sodium and potassium. Rubidium, caesium and francium are not available in schools because they are too reactive.

KNOWLEDGE CHECK

✓ Metals are positioned on the left-hand side and middle of the Periodic Table.
✓ Elements in a group have similar electron configuration.
✓ Metal oxides and hydroxides are basic and those that dissolve in water form alkalis.

LEARNING OBJECTIVES

✓ Describe the Group I alkali metals, lithium, sodium and potassium, as relatively soft metals with general trends down the group, limited to: decreasing melting point; increasing density; increasing reactivity with water.
✓ SUPPLEMENT Predict the properties of other elements in Group I, given information about the elements.

REACTIVITY OF GROUP 1 ELEMENTS

All Group I elements react with water to produce an alkaline solution. This makes them recognisable as a 'family' of elements, often called the **alkali metals.**

These very reactive metals all have only one electron in their outer electron shell. This electron is easily given away when the metal reacts with non-metals. The more electrons a metal atom has to lose in a reaction, the more energy is needed to start the reaction. This is why the Group II elements are less reactive – they have to lose two electrons when they react.

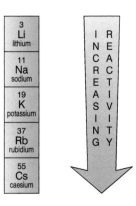

△ Fig. 8.7 Group I elements become more reactive as you go down the group.

GROUP I PROPERTIES

The properties of Group I metals are as follows:

- Soft to cut.
- Shiny when cut, but quickly tarnish in the air.
- Very low melting points compared with most metals – melting points decrease down the group.
- React very easily with water. Very low densities compared with most metals. Lithium, sodium and potassium float on water. Densities increase down the group.
- React very easily with air and water. The alkali metals are so reactive that they are stored in oil to prevent reaction with air and water. Reactivity increases down the group.

△ Fig. 8.8 The freshly cut surface of sodium.

Reaction	Observations	Equations
Air or oxygen	The metals burn easily and their compounds colour flames: lithium – red sodium – orange/yellow potassium – lilac A white solid oxide is formed	lithium + oxygen → lithium oxide $4Li(s) + O_2(g) \rightarrow 2Li_2O(s)$ sodium + oxygen → sodium oxide $4Na(s) + O_2(g) \rightarrow 2Na_2O(s)$ potassium + oxygen → potassium oxide $4K(s) + O_2(g) \rightarrow 2K_2O(s)$
Water	The metals react vigorously They float on the surface, moving around rapidly With both sodium and potassium, the heat of the reaction melts the metal so it forms a sphere; bubbles of gas are given off and the metal 'disappears' With the more reactive metals (such as potassium) the hydrogen gas produced burns The resulting solution is alkaline	lithium + water → lithium hydroxide + hydrogen $2Li(s) + 2H_2O(l) \rightarrow$ $2LiOH(aq) + H_2(g)$ sodium + water → sodium hydroxide + hydrogen $2Na(s) + 2H_2O(l) \rightarrow$ $2NaOH(aq) + H_2(g)$ potassium + water → potassium hydroxide + hydrogen $2K(s) + 2H_2O(l) \rightarrow$ $2KOH(aq) + H_2(g)$

△ Table 8.1 Reactions of Group I metals.

It is possible to predict information about the Group 1 elements using information about the trend in physical or chemical properties. For example, look at the following melting point information:

Group 1 element	Melting point /°C
lithium	181
sodium	98
potassium	64
rubidium	
caesium	28

The overall trend in melting points is clearly a decrease down the group. It is possible to make a prediction about the likely melting point of rubidium. The drop in melting points between one element and the next is not the same in each case. It is very likely to be between 64°C and 28°C, slightly below the mid-point of 46°C (the melting point of sodium is below the mid-point between the melting points of lithium and potassium). The melting point of rubidium is actually 40°C.

QUESTIONS

1. Explain why the Group I elements are known as the *alkali metals*.

2. Suggest how many electrons the Group I elements' atoms have in their outer shell.

3. The Group I metals are unusual metals. Give one property they have that is different to most other metals.

4. SUPPLEMENT Describe how the melting point of potassium will compare to that of sodium. Explain your answer.

5. SUPPLEMENT Table 8.2 shows data about Group I metals.
   The data for rubidium is missing.

Element	Melting point /°C	Rate of reaction with water
lithium	181	slow
sodium	98	moderate
potassium	64	fast
rubidium		

△ Table 8.2

a) Suggest the melting point of rubidium.

b) Predict the rate of reaction of rubidium with water.

COMPOUNDS OF THE GROUP I METALS

The compounds of Group I metals are usually colourless crystals or white solids and always have ionic bonding. Most of them are soluble in water. Some examples are sodium chloride (NaCl) and potassium nitrate (KNO_3).

The compounds of the alkali metals are widely used:

- lithium carbonate – as a hardener in glass and ceramics

- lithium hydroxide – removes carbon dioxide in air-conditioning systems

- sodium chloride – table salt

- sodium carbonate – a water softener

- sodium hydroxide – used in paper manufacture

- monosodium glutamate – a flavour enhancer

- sodium sulfite – a preservative

- potassium nitrate – a fertiliser; also used in explosives.

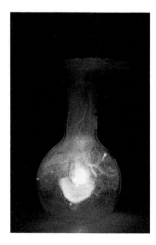

Δ Fig. 8.9 Sodium burning in chlorine.

Challenge Question: Give the chemical formulas for lithium carbonate and lithium hydroxide.

QUESTIONS

1. Suggest what gas is produced when potassium reacts with water. Give the name of the solution formed in this reaction.

2. **a)** Describe what a period is in the Periodic Table.

   b) Suggest which period the element magnesium is in.

   c) Explain in terms of electronic structures the difference in reactivity between sodium and argon.

3. These questions are based on Group I of the Periodic Table.

   a) Explain the term 'group' in the Periodic Table.

   b) Give the names of the first three metals of Group I.

   c) Explain why the elements in this group have similar chemical properties.

SCIENCE IN CONTEXT   FACTS ABOUT THE GROUP I METALS

1. Lithium is found in large quantities (estimated at 230 billion tonnes) in compounds in seawater.

2. Sodium is found in many minerals and is the sixth most abundant element overall in the Earth's crust (amounting to 2.6% by weight).

△ Fig. 8.10 Lithium is used in all of these batteries.

3. Potassium is also found in many minerals and is the seventh most abundant element in the Earth's crust (amounting to 1.5% by weight).

4. Rubidium was discovered by Bunsen (of Bunsen burner fame) in 1861. It is the 16th most abundant element and is more abundant than copper, about the same as zinc, and is found in very small quantities in a large number of minerals. Because of this low concentration in mineral deposits, only 2 to 4 tonnes of rubidium are produced each year worldwide.

5. Caesium is more abundant than tin, mercury and silver. However, its very high reactivity makes it very difficult to extract from mineral deposits.

6. Francium was discovered as recently as 1939 as a product of the radioactive decay of an isotope of actinium.

Challenge Question: Predict how the properties (for example, melting point, density and reactivity) of caesium compare to those of sodium.

End of topic checklist

Key terms

alkali metal, trend

During your study of this topic you should have learned:

○ How to describe lithium, sodium and potassium in Group I as a collection of relatively soft metals showing trends in melting point, density and reactivity with water.

○ **SUPPLEMENT** How to predict the properties of other elements in Group I, given information where appropriate.

End of topic questions

1. State which of these descriptions of sodium is NOT correct.

   A It is stored under water to prevent its reaction with the air.

   B It is soft and can be easily cut with a knife.

   C It is a solid with a low melting point.

   D It burns in air with an orange/yellow flame, forming sodium oxide.

2. This question is about the Group I elements lithium, sodium and potassium.

   a) State which is the most reactive of these elements.

   b) Explain why the elements are stored in oil.

   c) Explain why sodium floats when added to water.

3. Explain why the Group I elements are known as the 'alkali metals'.

4. Write word equations and balanced equations for the following reactions:

   a) lithium and oxygen

   b) potassium and water.

5. **SUPPLEMENT** This question is about rubidium (symbol Rb), which is a less common Group I element.

   a) Explain what state of matter would you predict rubidium to be in at room temperature and pressure.

   b) When rubidium is added to water:

      i) State which gas is formed.

      ii) Predict what chemical compound would be formed in solution. Predict what result would be seen if universal indicator was added to the solution.

   c) Would you expect rubidium to be more or less reactive than potassium? Justify your answer.

6. **SUPPLEMENT** Suggest why potassium is more reactive than sodium.

Group VII elements

INTRODUCTION

Group VII elements are located on the right-hand side of the Periodic Table with the other non-metals. They look very different from each other, so it may seem strange that they are in the same group. However, their chemical properties are very similar, and all of them are highly reactive. This topic focuses on chlorine, bromine and iodine. Fluorine is a highly reactive gas and astatine is a radioactive black solid with a very short half-life (so will exist in only very small quantities).

△ Fig. 8.11 At room temperature and atmospheric pressure, chlorine is a pale yellow-green gas, bromine a red-brown liquid and iodine is a grey-black solid.

KNOWLEDGE CHECK

✓ Non-metals are positioned on the right-hand side of the Periodic Table.
✓ Elements in a group have similar electronic configuration.
✓ Oxidation and reduction are important processes.

LEARNING OBJECTIVES

✓ Describe the Group VII halogens, chlorine, bromine and iodine, as diatomic non-metals with general trends down the group, limited to: increasing density; decreasing reactivity.
✓ State the appearance of the halogens at r.t.p. as: chlorine, a pale yellow-green gas; bromine, a red-brown liquid; iodine, a grey-black solid.
✓ **SUPPLEMENT** Describe and explain the displacement reactions of halogens with other halide ions.
✓ **SUPPLEMENT** Predict the properties of other elements in Group VII, given information about the elements.

REACTIVITY OF GROUP VII ELEMENTS

The Group VII elements are sometimes referred to as the **halogen** elements or halogens.

'Halogen' means 'salt-maker' – halogens react with most metals to make salts.

Halogen atoms have seven electrons in their outermost electron shell, so they need to gain only one electron to obtain a full outer shell.

This is what makes them very reactive. They react with metals, gaining an electron and forming a singly charged negative ion.

The reactivity of the elements decreases down the group.

GROUP VII PROPERTIES

The properties of the Group VII elements are as follows:

- The density of the elements increases down the group.
- Fluorine is a pale yellow gas; chlorine is a pale yellow-green gas; bromine is a red-brown liquid; iodine is a grey-black shiny solid.
- All the atoms have seven electrons in their outermost electron shell.
- All exist as diatomic molecules – each molecule contains two atoms. For example, F_2, Cl_2, Br_2, I_2.
- **SUPPLEMENT** They undergo **displacement reactions**.

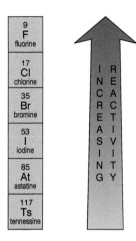

△ Fig. 8.12 Increasing reactivity up Group VII.

Reaction	Observations	Equations
SUPPLEMENT Displacement chlorine gas potassium iodide solution iodine being formed	A more reactive halogen will displace a less reactive halogen from a solution of a salt. Chlorine displaces bromine from sodium bromide solution. The colourless solution (sodium bromide) turns orange when chlorine is added due to the formation of bromine.	chlorine + sodium bromide → sodium chloride + bromine $Cl_2(g) + 2NaBr(aq) \rightarrow$ $2NaCl(aq) + Br_2(aq)$
	Chlorine displaces iodine from sodium iodide solution. The colourless solution (sodium iodide) turns brown when chlorine is added due to the formation of iodine.	chlorine + sodium iodide → sodium chloride + iodine $Cl_2(g) + 2NaI(aq) \rightarrow$ $2NaCl(aq) + I_2(aq)$

△ Table 8.3 Displacement reactions of the halogens.

SUPPLEMENT

In the displacement reactions between halogens and solutions of halide ions (shown above) the sodium ions (Na+) play no part in the reaction, so if the reaction between chlorine and sodium bromide solution is written with them removed we can write an **ionic equation**:

$$Cl_2(aq) + 2Br^-(aq) \rightarrow 2Cl^-(aq) + Br_2(aq)$$

This reaction can be written as two half-equations showing how electrons are involved:

$$Cl_2(aq) + 2e^- \rightarrow 2Cl^-(aq)$$

The chlorine has gained electrons – it has been reduced.

$$2Br^-(aq) \rightarrow Br_2(aq) + 2e^-$$

The bromide ions have lost electrons – they have been oxidised.

In this reaction the chlorine is an oxidising agent – it oxidises the bromide ions and is itself reduced. The bromide ions are reducing agents – they reduce the chlorine and are themselves oxidised.

QUESTIONS

1. Explain how many electrons the Group VII element atoms have in their outer shell.

2. Suggest why the Group VII elements are particularly reactive when compared to other non-metals.

3. Chlorine exists as diatomic molecules. Explain what this means.

4. SUPPLEMENT Astatine is an element in Group VII. Predict whether you would expect it to be a solid, liquid or gas at room temperature. Explain your answer.

5. SUPPLEMENT Explain what a 'displacement reaction' is, involving the Group VII elements.

6. SUPPLEMENT Bromine displaces iodine from potassium iodide solution. Write an ionic equation for this reaction.

SUPPLEMENT

Developing practical skills

A student was provided with three aqueous solutions containing chlorine, bromine and iodine. She added a few drops of a non-polar solvent to each solution in separate test-tubes and then stirred each with a clean glass rod. A non-polar solvent forms a separate layer when in aqueous solution. It floated on top of the student's solutions. When the non-polar solvent layer separated from each solution, she recorded the colours of the non-polar solvent layers in the table:

Solution in water	Colour of the non-polar solvent layer
chlorine	colourless
bromine	orange
iodine	violet

She cleaned out the tubes and then performed a series of test-tube reactions as indicated in the table below. In each case she mixed small quantities of solution A with twice the volume of solution B, added 10 drops of the non-polar solvent and then stirred the mixture with a clean glass rod. Once the layers had separated, she recorded her results.

Solution A	Solution B	Colour of non-polar solvent layer
aqueous chlorine	sodium bromide	orange
aqueous chlorine	sodium iodide	violet
aqueous bromine	sodium chloride	orange
aqueous bromine	sodium iodide	violet
aqueous iodine	sodium chloride	violet
aqueous iodine	sodium bromide	orange

Safety note: Eye protection and gloves should be worn when using or being exposed to chlorine or bromine.

Using and organising techniques, apparatus and materials

1. The non-polar solvent is highly flammable and toxic and the chlorine and bromine solutions are both irritants. Describe the precautions the student should have taken when doing this experiment.

Interpreting observations and data

2. What can you deduce about the relative reactivity of chlorine, bromine and iodine from the first two results in the second table?

3. Write an equation for the reaction indicated by result four in the second table.

Evaluating methods

4. The student made a mistake in recording one of her results. Identify which one. Explain how you know.

SCIENCE IN CONTEXT **USES OF HALOGENS**

Halogens and their compounds have a wide range of uses:

- fluorides – in toothpaste help prevent tooth decay

- fluorine compounds – making plastics like the non-stick surface on pans

- chlorofluorocarbons – propellants in aerosols and refrigerants (now being replaced because of their damaging effect on the ozone layer)

- chlorine – purifying water

- chlorine compounds – household bleaches

- hydrochloric acid – widely used in industry

- bromine compounds – making pesticides

- silver bromide – the light-sensitive film coating on photographic film

- iodine solution – an antiseptic.

FLUORINE

Fluorine is the most reactive non-metal in the Periodic Table. It reacts with most other elements except helium, neon and argon. These reactions are often sudden or explosive. Even radon, a very unreactive noble gas, burns with a bright flame in a jet of fluorine gas. All metals react with fluorine to form fluorides. The reactions of fluorine with Group I metals are explosive.

Early scientists tried to make fluorine from hydrofluoric acid (HF(aq)) but this proved to be highly dangerous, killing or blinding several scientists who attempted it. They became known as the 'fluorine martyrs'. Today fluorine is manufactured by the electrolysis of the mineral fluorite, which is calcium fluoride.

Fluorine is not an element to play with. You will certainly not see it in your laboratory!

Challenge Question: This question is about calcium fluoride.

a) Determine the chemical formula of calcium fluoride.

b) State the charges on the calcium and fluoride ions.

c) In the electrolysis of molten calcium fluoride, state at which electrode the fluorine would be produced.

QUESTIONS

You should be familiar with the elements of Group VII, the halogens. These are coloured non-metallic elements of varying reactivity. Although they are potentially harmful, their properties make them very useful. Use your knowledge of atomic structure, bonding and reaction types to answer the questions below.

1. SUPPLEMENT Chlorine is a pale yellow-green gas which can be obtained by the electrolysis of an aqueous solution of sodium chloride. Chlorine can be used to kill bacteria and is used in the manufacture of bleach.

   a) The electronic configuration of a chlorine atom is 2,8,7. Draw simple diagrams to show the arrangement of the outer electrons in a diatomic molecule of Cl_2 and a chloride ion, Cl^-.

   b) In the electrolysis of an aqueous solution of sodium chloride, the positive anode attracts the Cl^- ions to form chlorine molecules. Copy and complete the equation below and explain why this is oxidation.

   $$\underline{\hspace{3cm}} Cl^-(aq) \rightarrow \underline{\hspace{2cm}} (g) + 2e^-$$

c) Chlorine will displace bromine from a solution of potassium bromide to form bromine and potassium chloride. Explain why this reaction takes place and describe what you would observe if chlorine water was added to a solution of potassium bromide in a test-tube.

2. SUPPLEMENT Fluorine is a pale yellow gas and is the most reactive of the chemical elements. It is so reactive that glass, metals and even water burn with a bright flame in a jet of fluorine gas. Fluorides, however, are often added to toothpaste and, controversially, to some water supplies to prevent dental decay.

a) Write an ionic half-equation that shows the conversion of fluorine to fluoride ions.

b) Potassium fluoride is a compound that may be found in toothpaste. Explain why fluorine cannot be displaced from this compound using either chlorine or iodine.

End of topic checklist

Key terms

diatomic, halogens, trend, monatomic

SUPPLEMENT displacement reaction, ionic equation

During your study of this topic you should have learned:

○ How to describe chlorine, bromine and iodine in Group VII as a collection of diatomic non-metals showing trends in appearance, density and reactivity.

○ **SUPPLEMENT** How to describe and explain the displacement reactions of halogens with other halide ions.

○ **SUPPLEMENT** How to predict the properties of other elements in Group VII, given information about the elements.

End of topic questions

1. This question is about the Group VII elements: chlorine, bromine and iodine.

    a) State which is the most reactive of these elements.

    b) State which of the elements exists as a liquid at room temperature and pressure.

    c) State which of the elements exists as a solid at room temperature and pressure.

    d) Describe the appearance of bromine.

2. Write word equations for the following reactions:

    a) sodium and chlorine

    b) potassium and bromine

    c) lithium and iodine.

3. **SUPPLEMENT** Aqueous bromine reacts with sodium iodide solution.

    a) State what type of chemical reaction this is.

    b) Write a balanced equation for the reaction.

    c) The reaction involves oxidation and reduction.

       i) Define *oxidation*. State what has been oxidised in this reaction.

       ii) Define *reduction*. State what has been reduced in this reaction.

4. **SUPPLEMENT** Determine which of these equations correctly shows a displacement reaction of the halogen elements.

    A $2NaCl + Br_2 \rightarrow 2NaBr + Cl_2$

    B $2NaCl + I_2 \rightarrow 2NaI + Cl_2$

    C $2NaI + Cl_2 \rightarrow 2NaCl + I_2$

    D $2NaBr + I_2 \rightarrow 2NaI + Br_2$

Transition metals and noble gases

INTRODUCTION

There are two other important 'families' of elements. The first are the transition elements, which are a 'block' of metals that includes more 'everyday' metals than Group I.

The second is the noble gases (Group VIII), a group of elements of interest because of their uses rather than their chemical reactions.

△ Fig. 8.13 This incandescent light bulb contains unreactive argon instead of air.

KNOWLEDGE CHECK

✓ Metals are positioned on the left side and the middle of the Periodic Table.
✓ Non-metals are positioned on the right side of the Periodic Table.
✓ Elements in a group have similar electronic configuration.

LEARNING OBJECTIVES

✓ Describe the transition elements as metals that: have high densities; have high melting points; form coloured compounds; often act as catalysts as elements and in compounds.
✓ Describe the Group VIII noble gases as unreactive, monatomic gases and explain this in terms of electronic configuration.

TRANSITION ELEMENTS

The **transition metals** are grouped in the centre of the Periodic Table and include iron, copper, zinc and chromium.

All the transition metals have more than one electron in their outer electron shell. They are much less reactive than Group I and Group II metals and so are more 'everyday' metals. They have much higher melting points and densities. They react much more slowly with water and with oxygen.

They are widely used as construction metals (particularly iron through steel).

One of the typical properties of transition metals and their compounds is their ability to act as **catalysts** and speed up the rate of a chemical reaction.

Property	Group I metal	Transition metal
melting point	low	high
density	low	high
colour of compounds	white	mainly coloured
reactions with water/air	vigorous	slow or no reaction
reactions with an acid	violent (dangerous)	slow or no reaction

△ Table 8.4 Properties of the Group I metals and the transition metals.

You saw in Chapter 6 that oxidation numbers are used to name some ions, for example, iron(II) and iron(III). This is because many transition metals have more than one oxidation number. For example, copper forms Cu^+ ions and Cu^{2+} ions, as shown in its compounds – for example copper(I) oxide, Cu_2O, and copper(II) oxide, CuO.

Also, iron forms Fe^{2+} ions and Fe^{3+} ions, as shown in iron(II) hydroxide, $Fe(OH)_2$, and iron(III) hydroxide, $Fe(OH)_3$.

The compounds of the transition metals are usually coloured. Copper compounds are usually blue or green; iron compounds tend to be either green or brown. When sodium hydroxide solution is added to a solution of a transition metal compound, a precipitate of the metal hydroxide is formed. The colour of the precipitate helps to identify the metal, as you will learn in Section 12. For example:

copper(II) sulfate + sodium hydroxide → copper(II) hydroxide + sodium sulfate

$CuSO_4(aq)$ + $2NaOH(aq)$ → $Cu(OH)_2(s)$ + $Na_2SO_4(aq)$

SUPPLEMENT

This can be written as an ionic equation:

$Cu^{2+}(aq) + 2OH^-(aq) \rightarrow Cu(OH)_2(s)$

Colour of metal hydroxide	Likely metal present
blue	copper(II) Cu^{2+}
green turning to brown	iron(II) Fe^{2+}
orange/brown	iron(III) Fe^{3+}

△ Table 8.5 Transition metal hydroxides and their colours.

QUESTIONS

1. Answer these questions about zinc.

a) What is its proton number?

b) Describe what information the proton number gives about the structure of a zinc atom.

c) State which period of the Periodic Table zinc is in.

d) State whether zinc is a metal or a non-metal.

2. Explain whether you would expect a reaction to happen between copper and water.

3. Chromium forms a compound called chromium(III) oxide. Explain what the number in the name indicates.

4. a) Write a fully balanced equation for the reaction between iron(II) sulfate solution and sodium hydroxide solution.

b) Describe the colour of the precipitate formed.

NOBLE GASES

This is actually a group of very unreactive non-metals. They used to be called the inert gases as it was thought that they didn't react with anything. But scientists later managed to produce fluorine compounds of some of the **noble gases**. As far as your school laboratory work is concerned, however, they are completely unreactive.

Name	Symbol
Helium	He
Neon	Ne
Argon	Ar
Krypton	Kr
Xenon	Xe
Radon	Rn

△ Table 8.6 The noble gases.

The unreactivity of the noble gases can be explained in terms of their **electronic configurations**. The atoms all have complete outer electron shells or eight electrons in their outer shell. They don't need to lose electrons (as metals do), or gain electrons (as most non-metals do).

Similarities of the noble gases

- Full outer electron shells
- Very unreactive
- Gases
- Exist as single atoms – they are **monatomic** (He, Ne, Ar, Kr, Xe, Rn).

How are the noble gases used?

- Helium – in balloons
- Neon – in red tube lights
- Argon – in lamps and light bulbs.

QUESTIONS

1. Why is helium unreactive?

2. Neon is monatomic. What does this mean?

3. Find argon on the Periodic Table.

a) What is its proton number?

b) What is its period number, and what does this tell you about the number of occupied electron shells?

c) How many electrons does argon have in its outer electron shell?

d) Write the electronic configuration of argon.

End of topic checklist

Key terms

catalyst, electronic configuration, monatomic, noble gas, transition metal

During your study of this topic you should have learned:

○ How to describe the transition elements as a collection of metals with high densities and high melting points, forming coloured compounds and often acting as catalysts.

○ That the noble gases are unreactive, monatomic gases as result of their electronic configurations.

End of topic questions

1. State which of the following is a typical property of a transition element:

   A low density

   B high melting point

   C form white compounds

   D react vigorously with cold water.

2. This question is about the transition elements.

   a) Give two differences in the physical properties of the transition metals compared with the alkali metals.

   b) Transition metals are used as catalysts. Define the term 'catalyst'.

   c) Suggest why the alkali metals are more reactive than the transition metals.

3. Analyse the table of observations.

Compound tested	Colour of compound	Effect of adding sodium hydroxide solution to a solution of the compound
A	white	no change
B	blue	blue precipitate formed
C	white	white precipitate formed

   a) State which of the compounds, A, B or C, contains a transition metal. Explain your answer.

   b) State which transition metal you think it is.

   c) SUPPLEMENT Compound B is a metal sulfate. Write a balanced equation for the reaction between a solution of this compound and sodium hydroxide solution.

4. SUPPLEMENT Iron forms two hydroxides, iron(II) hydroxide and iron(III) hydroxide.

   a) Explain what the Roman numerals indicate about the iron in these compounds.

   b) Write the chemical formulas of these two hydroxides.

5. Explain why the noble gases are so unreactive.

6. The noble gases are monatomic. Describe what this means.

7. Although the noble gases are generally very unreactive, reactions do occur with very reactive elements such as fluorine. State which of the noble gases are more likely to react – helium at the top of the group or xenon near the bottom of the group.

In this section you will extend your learning of metallic bonding from Section 2, Atoms, elements and compounds, to the properties of metals, their uses and the importance of alloys. There are also strong links to Section 8, The Periodic Table, and the position of metals in the Periodic Table. You will then consider the difference in reactivity of metals and how this is linked to the extent to which they corrode. Focusing on the reactivity series is useful in summarising the difference in the reactivity of metals and how this influences the methods that can be used to extract metals from their ores. Specifically, the method used for the extraction of iron from its ore is compared to the method used for the extracting aluminium. In the case of the extraction of aluminium there are links to Section 4, Electrochemistry.

Starting points

1. Can you name some of the metals commonly used in construction? Where in the Periodic Table are these metals positioned?

2. What is the difference between iron and steel?

3. Can you name a metal that corrodes easily and one that resists corrosion?

4. What is a redox reaction?

5. What type of compounds behave as electrolytes?

SYLLABUS SECTIONS COVERED

9.1 Properties of metals

9.2 Uses of metals

9.3 Alloys and their properties

9.4 Reactivity series

9.5 Corrosion of metals

9.6 Extraction of metals

9
Metals

Δ The Atomium was built in 1958 using mainly aluminium sheets. These have now been replaced by stainless steel.

Metals

Δ Fig. 9.1 What sort of properties should the metals used in the construction of this helicopter have?

INTRODUCTION

Metals are very important in our everyday lives and many have very similar physical properties. Some metals are highly reactive, like the Group I metals on the left-hand side of the Periodic Table. Other metals are much less reactive, such as the transition metals in the middle of the Periodic Table. Knowing the order of the reactivity of metals can help chemists make very accurate predictions about how the metals will react with different substances and also what individual metals can be used for.

KNOWLEDGE CHECK

✓ Metals with different properties are found in different parts of the Periodic Table.
✓ Metals have different reactivities.
✓ Metals are very useful materials, often used in construction.

LEARNING OBJECTIVES

✓ Compare the general physical properties of metals and non-metals, including: thermal conductivity; electrical conductivity; malleability; melting and boiling points.
✓ Describe the general chemical properties of metals, limited to their reactions with: dilute acids; cold water and steam.
✓ Describe the uses of metals in terms of their physical properties, including:
　(a) aluminium in the manufacture of aircraft because of its low density
　(b) aluminium in the manufacture of overhead electrical cables because of its low density and good electrical conductivity
　(c) aluminium in food containers because of its resistance to corrosion
　(d) copper in electrical wiring because of its good electrical conductivity.
✓ Describe alloys as mixtures of a metal with other elements, including: brass as a mixture of copper and zinc; stainless steel as a mixture of iron and other elements, such as chromium, nickel and carbon.
✓ State that alloys can be harder and stronger than the pure metals and are more useful.
✓ Describe the uses of alloys in terms of their physical properties, including stainless steel in cutlery because of its hardness and resistance to rusting.
✓ Identify representations of alloys from diagrams of structure.
✓ **SUPPLEMENT** Explain in terms of structure how alloys can be harder and stronger than the pure metals because the different sized atoms in alloys mean the layers can no longer slide over each other.
✓ State the order of the reactivity series as: potassium, sodium, calcium, magnesium, aluminium, carbon, zinc, iron, hydrogen, copper, silver, gold.
✓ Describe the reactions, if any, of: potassium, sodium and calcium with cold water; magnesium with steam; magnesium, zinc, iron, copper, silver and gold with dilute hydrochloric acid, and explain these reactions in terms of the position of the metals in the reactivity series.
✓ Deduce an order of reactivity from a given set of experimental results.

- ✓ **SUPPLEMENT** Describe the relative reactivities of metals in terms of their tendency to form positive ions, by displacement reactions, if any, with the aqueous ions of magnesium, zinc, iron, copper and silver.
- ✓ State the conditions required for the rusting of iron (presence of oxygen and water).
- ✓ State some common barrier methods, including painting, greasing and coating with plastic.
- ✓ Describe how barrier methods prevent rusting by excluding oxygen or water.
- ✓ **SUPPLEMENT** Describe the use of zinc in galvanising as an example of a barrier method and sacrificial protection.
- ✓ **SUPPLEMENT** Explain sacrificial protection in terms of the reactivity series and in terms of electron loss.
- ✓ Describe the ease in obtaining metals from their ores, related to the position of the metal in the reactivity series.
- ✓ State that iron from hematite is extracted by reduction of iron (III) oxide in the blast furnace.
- ✓ **SUPPLEMENT** Describe the extraction of iron from hematite in the blast furnace, limited to:
    (a) the burning of carbon (coke) to provide heat and produce carbon dioxide $C + O_2 \rightarrow CO_2$
    (b) the reduction of carbon dioxide to carbon monoxide $C + CO_2 \rightarrow 2CO$
    (c) the reduction of iron(III) oxide by carbon monoxide $Fe_2O_3 + 3CO \rightarrow 2Fe + 3CO_2$
    (d) the thermal decomposition of calcium carbonate / limestone to produce calcium oxide $CaCO_3 \rightarrow CaO + CO_2$
    (e) the formation of slag $CaO + SiO_2 \rightarrow CaSiO_3$
- ✓ State that the main ore of aluminium is bauxite and that aluminium is extracted by electrolysis.

PROPERTIES OF METALS

Most metals have similar physical properties. These physical properties are very different from those of non-metals.

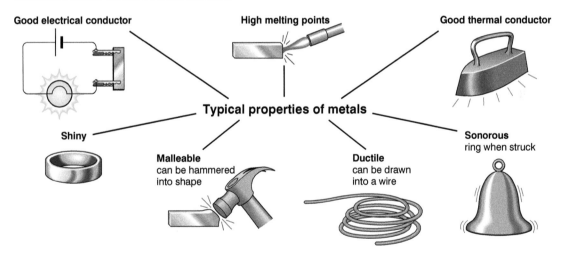

△ Fig. 9.2 Typical properties of metals.

Exceptions:

- The alkali metals have low melting points and are not sonorous.
- Mercury has a low melting point.

▷ Fig. 9.3 Some examples of non-metals. From left: silicon, chlorine, sulfur.

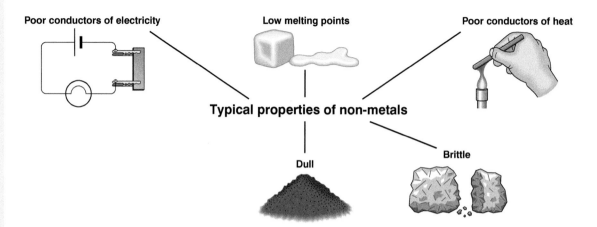

Poor conductors of electricity

Low melting points

Poor conductors of heat

Typical properties of non-metals

Dull

Brittle

Δ Fig. 9.4 Typical properties of non-metals.

Exceptions:
- Carbon in the form of graphite is a good conductor of electricity.
- Carbon and silicon have high melting points.

USES OF METALS

Uses of aluminium

The uses of aluminium are based on its properties of having a low density but being quite a strong metal as well as being unreactive (because of its oxide coating). These properties make it useful in the manufacture of aircraft (because it has a low density and is strong), overhead electrical cables (because of its low density and good electrical conductivity) and food containers (because it resists **corrosion**).

Uses of copper

Copper is an excellent conductor of electricity and is ductile. These properties make it an ideal material for use as electrical wires.

Copper is also an excellent conductor of heat with a high melting point. These properties make it ideal for cooking utensils.

ALLOYS AND THEIR PROPERTIES

An **alloy** is a mixture of a metal with one or more other elements.

The reason for producing alloys is to 'improve' the properties of a metal.

Δ Fig. 9.5 Alloys are used to make coins.

For example, stainless steel is used to make cutlery because of its hardness and resistance to corrosion.

Alloy	Constituents
Brass	Copper (70%), zinc (30%)
Stainless steel	Iron and other elements such as chromium, nickel and carbon

Δ Table 9.1 Common alloys.

The structure of alloys

The structure of pure metallic elements is usually shown as in Fig. 9.6.

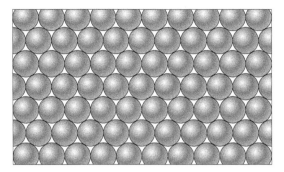

◁ Fig. 9.6 Particles in a solid.

This is a simplified picture but, surprisingly, such a structure is very weak. If there is the slightest difference between the planes of atoms, the metal will break at that point.

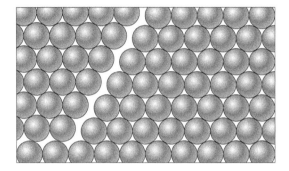

◁ Fig. 9.7 The gaps show a weak point of a metal.

The more irregular (jumbled-up) the metal atoms are, the stronger the metal is. This is why alloy structures are stronger: because of the elements added.

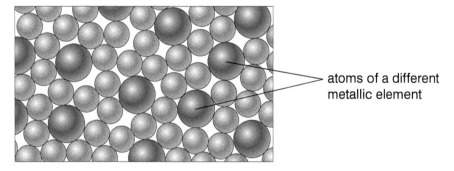

atoms of a different metallic element

Δ Fig. 9.8 Atoms in an alloy.

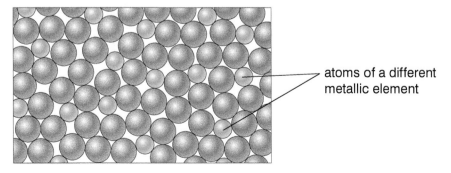

atoms of a different metallic element

Δ Fig. 9.9 Smaller atoms also make the metal stronger.

Steel that is heated to red heat and then plunged into cold water is made harder by the process of 'jumbling up' the metal atoms. Further heat treatment is used to increase the strength and toughness of the alloy.

SUPPLEMENT

The structure of an alloy can be used to explain why the alloy is stronger than the pure metal. The different sized atoms that have been added to the pure metal prevent the layers of atoms from sliding over each other.

QUESTIONS

1. Metals are usually *malleable*. Explain the term 'malleable'.

2. Explain what an alloy is.

3. SUPPLEMENT Explain why an alloy of aluminium is likely to be stronger than pure aluminium.

REACTIVITY SERIES

The Periodic Table is a way of ordering the chemical elements that highlights their similar and different properties. The **reactivity series** is another way of classifying elements, this time in order of their reactivity, to help explain or predict their reactions. This has many practical applications, such as being able to predict how metals can be extracted from their ores.

The more reactive metals react with oxygen to form oxides:

calcium + oxygen → calcium oxide

$$2Ca(s) + O_2(g) → 2CaO(s)$$

Less reactive metals, such as gold, do not react with oxygen.

△ Fig. 9.10 Sodium, magnesium, gold – these are all metals, but they have very different reactivities.

Elements can be arranged in order of their reactivity. The more reactive a metal is, the easier it is to form compounds and the harder it is to break those compounds down. We can predict how metals might react by looking at the reactivity series (Fig. 9.11).

The most reactive metals react with water at room temperature. For example, potassium, sodium and lithium in Group I and calcium in Group II react rapidly with water:

sodium + water → sodium hydroxide + hydrogen

$2Na(s) + 2H_2O(l) → 2NaOH(aq) + H_2(g)$

The less reactive metals such as magnesium and iron react with steam:

magnesium + steam → magnesium oxide + hydrogen

$Mg(s) + H_2O(g) → MgO(s) + H_2(g)$

Some of the mid-reactivity series metals produce hydrogen when they react with dilute acids. So, for example, magnesium, aluminium, zinc and iron all release hydrogen when they react with dilute hydrochloric acid.

zinc + hydrochloric acid → zinc chloride + hydrogen

$Zn(s) + 2HCl(aq) → ZnCl_2(aq) + H_2(g)$

The metals below hydrogen in the reactivity series, such as silver and gold, do not react to form hydrogen with water or dilute acids.

Another use of the reactivity series is to predict how metals can be extracted from their ores. The elements below carbon in the reactivity series can be obtained by heating their oxides with carbon:

zinc oxide + carbon → zinc + carbon dioxide

$2ZnO(s) + C(s) → 2Zn(s) + CO_2(g)$

copper(II) oxide + carbon → copper + carbon dioxide

$2CuO(s) + C(s) → 2Cu(s) + CO_2(g)$

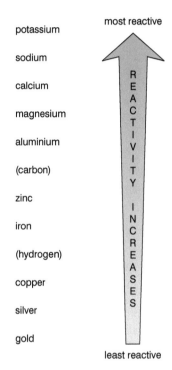

△ Fig. 9.11 The reactivity series shows elements, mainly metals, in order of decreasing reactivity.

SUPPLEMENT

This type of reaction is called a **displacement reaction**. A more reactive element, such as carbon, 'pushes' (or displaces) a less reactive metal, such as copper, out of its compound. In this reaction the copper(II) oxide has lost oxygen and been reduced. The carbon has gained oxygen and been oxidised.

The position of a metal in the reactivity series depends on how easily it forms positive ions. More reactive metals will form ions more readily than less reactive metals.

Any element higher up the reactivity series can displace an element lower down the series.

For example, magnesium is higher up the reactivity series than copper. So if magnesium powder is heated with copper(II) oxide, then copper and magnesium oxide are produced:

magnesium + copper(II) oxide → magnesium oxide + copper

$Mg(s)$ + $CuO(s)$ → $MgO(s)$ + $Cu(s)$

This reaction is an example of a redox reaction. The magnesium has been oxidised to magnesium oxide and the copper(II) oxide has been reduced to copper. Because the magnesium is responsible for the **reduction** of the copper(II) oxide, it is acting as a reducing agent. Similarly, the copper(II) oxide is responsible for the **oxidation** of the magnesium, so it is acting as an oxidising agent. In a redox reaction, the reducing agent is always oxidised and the oxidising agent is always reduced.

What will happen if copper is heated with magnesium oxide? Nothing happens, because copper is lower in the reactivity series than magnesium.

Using displacement reactions to establish a reactivity series

Displacement reactions of metals and their compounds in aqueous solution can be used to work out the order in the reactivity series.

In the same way that a more reactive element can push a less reactive element out of a compound, a more reactive metal ion in aqueous solution can displace a less reactive one.

For example, if you add zinc to copper(II) sulfate solution, the zinc displaces the copper because zinc is more reactive than copper. When the experiment is carried out, the blue colour of the copper(II) ion will fade as copper is produced and zinc ions are made:

zinc + copper(II) sulfate solution → zinc sulfate solution + copper

$Zn(s) + Cu^{2+}(aq) + SO_4^{2-}(aq)$ → $Zn^{2+}(aq) + SO_4^{2-}(aq) + Cu(s)$

To build up a whole reactivity series, a set of reactions can be tried to see if metals can displace other metal ions. By following the general rule that a more reactive metal can displace a less reactive metal it is possible to establish the reactivity series.

For example, you may have seen the reaction of copper wire with silver nitrate solution. As the reaction proceeds, a shiny grey precipitate appears (this is silver) and the solution begins to turn blue as Cu^{2+} ions are produced from the copper.

Copper + silver nitrate → copper(II) nitrate + silver

$Cu(s)$ + $2AgNO_3(aq)$ → $Cu(NO_3)_2(aq)$ + $2Ag(s)$

This shows that silver can be displaced by copper, and so silver is below copper in the reactivity series.

QUESTIONS

1. Determine whether copper will react with dilute hydrochloric acid to produce hydrogen. Explain your answer.

2. Plan and design an experiment to contrast the reactions of the metals aluminium, copper, iron, magnesium and zinc with dilute hydrochloric acid. You should:

   a) demonstrate how to safely use techniques, apparatus and materials

b) plan the procedure you will use

c) show how you will record your measurements or observations.

3. SUPPLEMENT Write a balanced equation for the reaction of potassium with water.

4. SUPPLEMENT Determine whether carbon will displace magnesium from magnesium oxide. Explain your answer.

5. SUPPLEMENT Write a balanced ionic equation for the reaction between magnesium and lead(II) oxide.

SUPPLEMENT

Developing practical skills

A student was asked to carry out some possible displacement reactions. She was given samples of four metals A, B, C and D and a solution of each of their metal nitrates. She set up a series of test-tube reactions as summarised in the table:

Solution	Metal A	Metal B	Metal C	Metal D
Metal A nitrate, $A(NO_3)_2$ (aq)		Yes	Yes	No
Metal B nitrate, $B(NO_3)_2$ (aq)	No		No	No
Metal C nitrate, $C(NO_3)_2$ (aq)	No	Yes		No
Metal D nitrate, $D(NO_3)_2$ (aq)	10	11	12	

She decided that she would need 12 test-tubes. In each test-tube she put a 1 cm depth of one of the solutions and then added a small piece of one of the metals. She wore her eye protection until she had finished and tried to avoid getting any of the chemicals on her skin. She left the tubes for 10 minutes and then examined the solution and the piece of metal to see if any reaction was evident. She then recorded a 'yes' if a displacement reaction had taken place and a 'no' where no reaction was evident. She didn't have time to record her results for the metal D nitrate solution (tubes 10, 11 and 12).

Using and organising techniques, apparatus and materials

1. Justify why the student didn't set up the tubes represented by the blank rectangles.

2. Even though the student didn't record her results for metal D nitrate solution, explain why she would still be able to put the metals in order of reactivity.

Interpreting observations and data

3. Use the results to deduce the order of reactivity of the four metals. Start with the most reactive metal.

4. Predict the results you would expect for the three reactions 10, 11 and 12.

5. Write a balanced equation for the displacement reaction between metal B and metal C nitrate solution. (Use the symbols B and C for the two metals.)

6. Metal D nitrate solution was blue and metal D was a shiny orange colour. Suggest a name for metal D.

CORROSION OF METALS

Rusting is a chemical reaction between iron, water and oxygen. Water and oxygen must both be present for rusting to occur. The process is speeded up if there are also electrolytes, such as salt, in the water. This is why rusting takes place much faster in sea water.

The rusting of iron and steel can be prevented in a number of ways. The use of grease, oil, paint or coating with plastic, the so-called **barrier methods**, prevent water and oxygen from reaching the metal surface. However, these coatings can be damaged and then rusting will take place.

SUPPLEMENT

Iron can be prevented from rusting by using what we know about the reactivity series. Zinc is above iron in the reactivity series; that is, zinc reacts more readily than iron.

Galvanised iron is iron that is coated with a layer of zinc. To begin with, the coating will protect the iron. If the coating is damaged or scratched, the iron is still protected from rusting. This is because zinc is more reactive than iron and so it reacts and corrodes instead of the iron.

If zinc blocks are attached to the hulls of ships, they will corrode instead of the hull. The zinc block provides **sacrificial protection**.

Δ Fig. 9.12 Galvanised iron or steel resists corrosion by air and water.

SCIENCE IN CONTEXT

IRON AND STEEL

In the presence of water and oxygen iron and steel will both rust. Rust is actually hydrated iron(III) oxide. Rusting can be a major problem.

Steel is iron with 0.1–1.5% carbon content. Steel is more resistant to corrosion and is less brittle than iron. It has a wide range of uses, depending on its carbon content. For example:

- low carbon (<0.3%): car bodies
- medium carbon (0.3–0.9%): railway tracks
- high carbon (0.9–1.5%): knives.

Stainless steels are made by adding a wide range of metals, such as chromium, nickel, vanadium and cobalt, to the steel. Each gives the steel particular properties for specific uses. For example, vanadium steel is used to make high-precision, hard-wearing industrial tools.

Challenge Question: Write an equation showing the formation of iron(III) oxide (rust) from iron and oxygen in the presence of water.

1. Give the conditions necessary for iron to rust.

2. What is one disadvantage of using grease to prevent the rusting of iron?

3. SUPPLEMENT a) Explain what galvanising is.

   b) Explain why galvanised iron or steel is still protected from corrosion even when its surface has been scratched.

EXTRACTION OF METALS

Metals are found in the form of **ores** containing **minerals** mixed with unwanted rock. In almost all cases, the mineral is a compound of the metal, not the pure metal. One exception is gold, which can exist naturally in a pure state.

Extracting a metal from its ore usually involves two steps:

1. The mineral is physically separated from unwanted rock.

2. The mineral is broken down chemically to obtain the metal.

Reactivity of metals

The chemical method chosen to break down a mineral depends on the reactivity of the metal. The more reactive a metal is, the harder it is to break down its compounds. The more reactive metals are obtained from their minerals by the process of **electrolysis**. For example, aluminium is obtained from its ore, bauxite, by electrolysis.

The less reactive metals can be obtained by heating their oxides with carbon. This method will only work for metals below carbon in the reactivity series. It involves the **reduction** of a metal oxide to the metal.

Metal		Extraction method
Potassium	}	The most reactive metals are obtained using electrolysis
Sodium		
Calcium		
Magnesium		
Aluminium		
(Carbon)		
Zinc	}	These metals are below carbon in the reactivity series and so can be obtained by heating their oxides with carbon
Iron		
Lead		
Copper		
Silver	}	The least reactive metals are found as pure elements
Gold		

△ Table 9.2 Methods for extracting different metals.

Extracting iron

Iron is produced on a very large scale by reduction using carbon. The reaction takes place in a huge furnace called a blast furnace.

Three important raw materials are put in the top of the furnace: iron ore (iron(III) oxide), coke (the source of carbon for the reduction) and limestone, to remove the impurities as slag. Iron ore is also known as hematite.

△ Fig. 9.13 Coke (nearly pure carbon).

△ Fig. 9.14 Iron ore (hematite).

△ Fig. 9.15 Limestone.

△ Fig. 9.16 Molten iron.

△ Fig. 9.17 Slag.

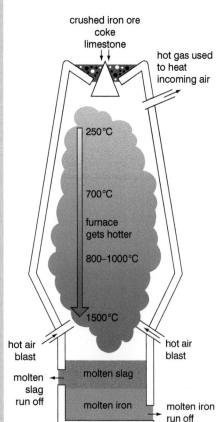

crushed iron ore
coke
limestone

hot gas used
to heat
incoming air

250°C

700°C

furnace
gets hotter

800–1000°C

1500°C

hot air
blast

hot air
blast

molten
slag
run off

molten slag

molten iron

molten iron
run off

1 Crushed iron ore, coke and limestone are fed into the top of the blast furnace

2 Hot air is blasted up the furnace from the bottom

3 Oxygen from the air reacts with coke to form carbon dioxide:
$$C(s) + O_2(g) \rightarrow CO_2(g)$$

4 Carbon dioxide reacts with more coke to form carbon monoxide:
$$CO_2(g) + C(s) \rightarrow 2CO(g)$$

5 Carbon monoxide is a reducing agent. Iron(III) oxide is reduced to iron:
⎡ reduction = loss of oxygen ⎤
$$Fe_2O_3(s) + 3CO(g) \rightarrow 2Fe(l) + 3CO_2(g)$$

6 Dense molten iron runs to the bottom of the furnace and is run off. There are many impurities in iron ore. The limestone helps to remove these as shown in processes 7 and 8.

7 Limestone is broken down by heat to calcium oxide:
$$CaCO_3(s) \rightarrow CaO(s) + CO_2(g)$$

8 Calcium oxide reacts with impurities like sand (silicon dioxide) to form a liquid called 'slag':
$$CaO(s) + SiO_2(s) \rightarrow CaSiO_3(l)$$
                impurity   slag
The liquid slag runs to the bottom of the furnace and is tapped off.

△ Fig. 9.18 How iron is extracted in a blast furnace.

The overall reaction is:

iron(III) oxide $+$ carbon $\rightarrow$ iron $+$ carbon dioxide

$2Fe_2O_3(s) \quad + \; 3C(s) \quad \rightarrow \; 4Fe(s) \; + \; 3CO_2(g)$

The reduction happens in three stages.

Stage 1 – The coke (carbon) reacts with oxygen 'blasted' into the furnace producing heat and carbon dioxide:

carbon $+$ oxygen $\rightarrow$ carbon dioxide

$C(s) \quad + \; O_2(g) \quad \rightarrow \; CO_2(g)$

Stage 2 – The carbon dioxide is reduced by unreacted coke to form carbon monoxide:

carbon dioxide $+$ carbon $\rightarrow$ carbon monoxide

$CO_2(g) \quad\quad + \; C(s) \quad\quad \rightarrow \; 2CO(g)$

Stage 3 – The iron(III) oxide is reduced by the carbon monoxide to iron:

iron(III) oxide $+$ carbon monoxide $\rightarrow$ iron $+$ carbon dioxide

$Fe_2O_3(s) \quad\quad + \; 3CO(g) \quad\quad\quad \rightarrow \; 2Fe(s) \; + \; 3CO_2(g)$

Stage 4 – The limestone (calcium carbonate) is thermally decomposed to calcium oxide.

$CaCO_3(s) \quad \rightarrow \quad CaO(s) \; + \; CO_2(g)$

Stage 5 – The calcium oxide reacts with sand (silicon(IV) oxide) to form slag which can be separated from the iron.

$CaO(s) \; + \; SiO_2(s) \; \rightarrow \; CaSiO_3(l)$

QUESTIONS

1. Describe which solid raw materials are used in the blast furnace.

2. The iron ore used in the blast furnace is usually hematite. Give the name of the main compound present in the ore.

3. Identify the gases that will escape from the top of the blast furnace.

4. SUPPLEMENT Write a balanced equation to show the reduction of iron(III) oxide by carbon.

5. This question is about steel.

   a) Steel is an alloy. Explain the term 'alloy'.

   b) Explain why is steel often used instead of the iron produced in the blast furnace.

The reactivity series of metals is useful to scientists in determining methods for metal extraction, ways of preventing corrosion and ensuring safe storage for reactive metals. Use the reactivity series to help answer the questions below.

6. Iron is a metal in common use, but unfortunately it rusts.

   a) Give the conditions needed for iron to rust.

   b) SUPPLEMENT Explain how galvanising prevents the rusting of iron.

RECYCLING

Extracting metals from their ores is an expensive process and so recycling metal objects can be economically worthwhile as well as environmentally more efficient. Recycling of metals essentially involves melting the metal and then using the molten metal to form a new object. Steel and aluminium are metals which are often recycled rather than being put in landfill sites. One potential problem is separating the different types of metal. For example, a motor car as well as being made of steel may include some parts made of aluminium (as well as a range of other materials). Separating the two different metals needs to be done before they can be recycled. In some cases recycling could prove more expensive than extracting the metal form its ore. However, the supplies of metal ores are limited and will eventually be used up. Recycling metals such as steel and aluminium will therefore become increasingly essential.

Challenge Question: Analyse the information provided. Do you think metals should be recycled even if this is a very expensive process? Give reasons to justify your answer.

THE EXTRACTION OF METALS

The reactivity of a metal determines how it can be extracted from ores from the Earth's crust. It also explains why some metals have been used for thousands of years while others have only been used much more recently.

The most unreactive metals can be found in their 'native' state, which is as the pure metal and not combined with other elements. Examples of such metals include gold and silver – metals that have been used for thousands of years. It is estimated that gold was first discovered in about 3000 BCE.

△ Fig. 9.19 This gold shoulder cape from North Wales is nearly 4000 years old and still in good condition.

Metals below carbon in the reactivity series can be extracted by heating their ores with carbon. Examples include lead and iron. It is possible that lead was discovered by accident when the silvery element was seen in the ashes of a wood fire that had been made above a deposit of lead ore. It is estimated that lead was first discovered in about 2000 BCE.

The most reactive metals, all those above carbon in the reactivity series, have to be extracted from their minerals by electrolysis. This process is a much more recent development and explains why these metals were not used until relatively recently. Aluminium was first extracted in 1825.

Challenge Question: Titanium can be found as an ore called rutile, which is titanium(IV) oxide.

a) State the chemical formula of rutile.

b) Explain why it is important to know whether titanium is more or less reactive than carbon.

End of topic checklist

Key terms

alloy, barrier methods, corrosion, electrical conductivity, electrolysis, malleability, mineral, ore, oxidation, reactivity series, reduction, rusting, thermal conductor

SUPPLEMENT displacement reaction, galvanising, sacrificial protection

During your study of this topic you should have learned:

○ How to compare the general physical properties of metals with non-metals.

○ How to describe the general chemical properties of metals (for example, reactions with dilute acids, cold water and steam).

○ That an alloy is a mixture of a metal with other elements (brass from copper and zinc, stainless steel from iron, carbon, chromium or nickel).

○ That metals are often used in the form of alloys as they can be harder and stronger than the pure metal and more resistant to rusting.

○ How to identify representations of an alloy from a diagram of its structure due to the impact of different sized atoms.

○ That the order of the reactivity series of metals is potassium, sodium, calcium, magnesium, aluminium (carbon), zinc, iron, (hydrogen), copper, silver and gold.

○ How to deduce an order of reactivity from a given set of experimental results.

○ How to describe the ease of obtaining metals from their ores by relating the metals to their positions in the reactivity series.

○ How to describe the essential reactions in the extraction of iron from hematite in a blast furnace.

○ That aluminium is extracted from the ore bauxite by electrolysis.

○ **SUPPLEMENT** How to explain in terms of structure how alloys can be harder or stronger than the pure metals.

○ **SUPPLEMENT** That relative reactivities of metals in the reactivity series are related to the tendency of a metal to form its positive ion, illustrated by its displacement reaction, if any, with the aqueous ions of other metals.

○ **SUPPLEMENT** The symbol equations for the extraction of iron from hematite.

○ That aluminium is used in the manufacture of aircraft because of its strength and low density.

○ That aluminium is used in food containers because of its resistance to corrosion.

○ That aluminium is used in the manufacture of overhead electrical cables because of its low density and good electrical conductivity.

○ That copper can be used for electrical wiring because of its good electrical conductivity.

○ That barrier methods such as painting, greasing and coating with plastic prevent rusting by excluding oxygen and water.

○ **SUPPLEMENT** That the use of zinc in galvanising is an example of a barrier method and sacrificial protection.

○ **SUPPLEMENT** How sacrificial protection is dependent on the reactivity of the metal.

1. Aluminium is often used instead of copper in overhead electrical cables. State which statement is NOT an important reason for choosing aluminium for this use.

   A Aluminium is a good electrical conductor.

   B Aluminium is a good thermal conductor.

   C Aluminium is easy to recycle.

   D Aluminium has a lower density.

2. Arrange the following metals in order of reactivity, starting with the most reactive: calcium, copper, magnesium, sodium, zinc.

3. **SUPPLEMENT** This question is about four metals represented by the letters Q, X, Y and Z. A series of displacement reactions was carried out and the results are shown below:

   Reaction 1: Q oxide + Y → Y oxide + Q

   Reaction 2: X oxide + Z → Z oxide + X

   Reaction 3: Q oxide + Z → no change

   a) Arrange the metals in order of reactivity, starting with the most reactive.

   b) In reaction 1:

   i) State which substance has been oxidised.

   ii) State which substance has been reduced.

   iii) State which substance is the oxidising agent.

   iv) State which substance is the reducing agent.

4. The least reactive metals such as gold and silver are found in their native state. Explain what you understand by this.

5. **SUPPLEMENT** Iron is extracted from iron ore (iron(III) oxide) in a blast furnace by heating with coke (carbon).

   a) Write a balanced equation including state symbols for the overall reaction.

   b) State whether the iron(III) oxide is oxidised or reduced in this reaction. Explain your answer.

   c) Explain why limestone is also added to the blast furnace.

6. Give reasons why:

a) Aluminium is used in the manufacture of aircraft.

b) Aluminium is used in the manufacture of food containers.

7. SUPPLEMENT Copper(II) sulfate solution reacts with zinc as shown below:

$$CuSO_4(aq) + Zn(s) \rightarrow ZnSO_4(aq) + Cu(s)$$

a) State what type of chemical reaction this is.

b) What can be deduced about the relative reactivities of copper and zinc?

8. SUPPLEMENT This question is about the reaction between magnesium and lead(II) oxide.

a) Write a balanced equation for the reaction.

b) State which is the more reactive metal, magnesium or lead.

c) This is an example of a redox reaction. Define the term 'redox'.

9. SUPPLEMENT Zinc can be extracted from zinc oxide by heating with carbon.

a) Write the balanced equation, including state symbols, for this reaction.

b) Zinc could also be extracted by the electrolysis of molten zinc oxide. Suggest why heating with carbon is the preferred method of extraction.

10. SUPPLEMENT Zinc prevents iron ships' hulls from rusting (corroding).

a) Give the name for this way of preventing rusting.

b) Explain how zinc protects iron from rusting.

c) Identify three other methods of preventing iron from rusting.

This section focuses on water and air, the key ingredients of the environment. You will learn about the chemical nature of the environmental issues facing the world today. Air quality, and the climate more generally, are also addressed, including the causes of global warming and the impact of acid rain. There are strong links to Section 11, Organic chemistry, through the issues related to the combustion of hydrocarbon, non-renewable fuels. In addition, the nature of some of the atmospheric pollutants will link to work in Section 7, Acids, bases and salts. There will be a strong focus on the strategies that can be used to reduce the effects on the climate of global warming and atmospheric pollution.

Starting points

1. Do you know the common test used to identify the presence of water?

2. Can you name some of the main gases in clean, dry air?

3. Can you name some of the gases that cause pollution in the air?

4. SUPPLEMENT Do you know about the greenhouse effect? If so, can you explain what it is?

SYLLABUS SECTIONS COVERED

10.1 Water

10.2 Air quality and climate

Chemistry of the environment

△ Water from rivers contains both solids, such as mud, and dissolved solids and microorganisms. Only distilled water contains nothing but water molecules.

Chemistry of the environment

INTRODUCTION

Clean water and air are precious. Water from natural sources is not a pure substance. It contains a variety of substances, some of which are beneficial and some are potentially harmful. Domestic water supplies have to be purified before they can be used safely.

Air also is not a pure substance. It provides the oxygen that all living things need to survive, and the carbon dioxide that plants need when they photosynthesise. Nitrogen

△ Fig. 10.1 The smog over the Forbidden City in Beijing is so thick that it obscures the view from Feng Shui Hill.

in the air is also very important for healthy plant growth, but not all plants can make use of nitrogen in this form, hence the need for fertilisers. Unfortunately not all the air we breathe is clean. It may contain a number of pollutants that can be harmful to living things and the environment. It is important to understand how these pollutants are produced and how they can be prevented from contaminating the air.

KNOWLEDGE CHECK

✓ Oxygen is present in the air and forms oxides when substances burn in it.
✓ Oxides of non-metals are acidic.
✓ Salts can be anhydrous or hydrated.

LEARNING OBJECTIVES

✓ Describe chemical tests for the presence of water using anhydrous cobalt(II) chloride and anhydrous copper(II) sulfate.
✓ Describe how to test for the purity of water using melting point and boiling point.
✓ State that distilled water is used in practical chemistry rather than tap water because it contains fewer chemical impurities.
✓ Describe the treatment of the domestic water supply in terms of: sedimentation and filtration to remove solids; use of carbon to remove tastes and odours; chlorination to kill microbes (pathogens).
✓ State the composition of clean, dry air as approximately 78% nitrogen, N_2, 21% oxygen, O_2, and the remainder as a mixture of noble gases and carbon dioxide, CO_2.
✓ State the source of each of these air pollutants, limited to:
  (a) carbon dioxide from the complete combustion of carbon-containing fuels
  (b) carbon monoxide and particulates from the incomplete combustion of carbon-containing fuels
  (c) methane from the decomposition of vegetation and waste gases from digestion in animals
  (d) oxides of nitrogen from car engines
  (e) sulfur dioxide from the combustion of fossil fuels which contain sulfur compounds.

✓ State the adverse effect of these air pollutants, limited to:
  (a) carbon dioxide: higher levels of carbon dioxide leading to increased global warming, which leads to climate change
  (b) carbon monoxide: toxic gas
  (c) particulates: increased risk of respiratory problems and cancer
  (d) methane: higher levels of methane leading to increased global warming, which leads to climate change
  (e) oxides of nitrogen: acid rain and respiratory problems
  (f) sulfur dioxide: acid rain.
✓ State and explain strategies to reduce the effects of climate change:
  (a) planting trees
  (b) reduction in livestock farming
  (c) decreasing use of fossil fuels
  (d) increasing use of hydrogen and renewable energy, e.g. wind, solar.
✓ **SUPPLEMENT** State and explain strategies to reduce the effects of acid rain: reducing emissions of sulfur dioxide by using low-sulfur fuels and flue gas desulfurisation with calcium oxide.
✓ **SUPPLEMENT** Describe how the greenhouse gases carbon dioxide and methane cause global warming, limited to: the absorption, reflection and emission of thermal energy; reducing thermal energy loss to space.
✓ **SUPPLEMENT** Explain how oxides of nitrogen form in car engines and describe their removal by catalytic converters, limited to $2CO + 2NO \rightarrow 2CO_2 + N_2$

WATER

The test for water is to add it to **anhydrous** copper(II) sulfate solid. If the liquid contains water, the powder will turn from white to blue as hydrated copper(II) sulfate forms.

The equation for the reaction is:

anhydrous copper(II) sulfate + water →
                    hydrated copper(II) sulfate

$$CuSO_4(s) + 5H_2O(l) \rightarrow CuSO_4 \bullet 5H_2O(s)$$

The presence of water can also be detected using anhydrous cobalt(II) chloride. The blue anhydrous cobalt(II) chloride turns to pink hydrated cobalt(II) chloride. A convenient way of performing the test is to use cobalt(II) chloride paper:

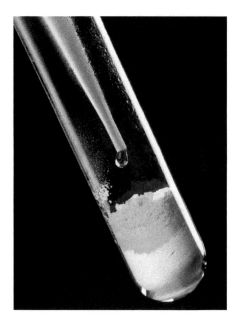

△ Fig. 10.2 Chemical test for water.

anhydrous cobalt(II) chloride + water →
                    hydrated cobalt(II) chloride

$$CoCl_2(s) + 6H_2O(l) \rightarrow CoCl_2 \bullet 6H_2O(s)$$

Neither of these tests shows the water is pure – only that the liquid has water in it.

Testing the purity of water

There are several ways to check the purity of water, but the simplest way is to heat the water to boiling point and measure the boiling point. Under normal atmospheric pressure pure water will boil at 100 °C. If the boiling point is lower or higher than this the water contains impurities. Another approach is to find the freezing point of the water, but this is not as simple to perform. Under normal atmospheric pressure pure water will freeze at 0 °C. Mixtures boil and freeze (or melt) over a range of temperatures.

You will be well aware that the water used in practical chemistry is distilled water. As the name suggests this is made by distilling a sample of water, possibly obtained from the domestic water supply. A simple distillation apparatus is shown in Fig. 12.12. The distilled water may not be absolutely pure but it will contain fewer chemical impurities.

The treatment of water supplies

Water is essential for life on Earth, and the demand for drinking water is increasing as the world's population grows. Two-thirds of the water is used in homes for washing, cleaning, cooking and in toilets. The rest is used by industry. Most industrial processes use water either as a raw material or for cooling. For example, it takes 200 000 litres of water to make 1 tonne of steel.

Water stored in reservoirs must be purified to produce drinkable tap water.

Water from reservoirs goes to a water treatment plant. Some sediments will settle to the bottom of the reservoir.	▶	Water is filtered through coarse gravel to remove larger pieces of dirt.	▶	Water is filtered through beds of fine gravel and sand to remove small particles.	▶	Chlorine is passed through to kill bacteria.	▶	Water supply to homes and industry.

In addition, tap water in certain areas is treated with carbon to remove tastes and odours.

In certain parts of the world supplies of water are very limited. There is insufficient clean water to drink and not enough water to irrigate and support the growth of crops. Water that is not purified can often contain harmful bacteria and so is not safe to drink. In the absence of clean water people have little choice but to drink contaminated water and risk illness or death.

AIR QUALITY AND CLIMATE

Clean air is a mixture of gases. Its current composition is shown in Fig. 10.3. There are concerns about the effect of the increasing levels of carbon dioxide which is linked to increased **global warming**.

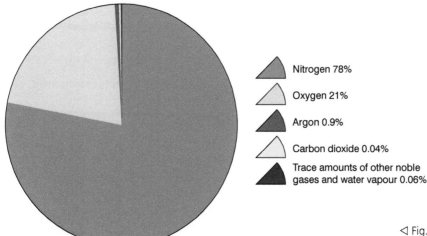

Nitrogen 78%

Oxygen 21%

Argon 0.9%

Carbon dioxide 0.04%

Trace amounts of other noble gases and water vapour 0.06%

◁ Fig. 10.3 Components of air.

Developing practical skills

A student set up the apparatus as shown in Fig. 10.4 with the long tube turned upside-down in a trough of water. Previously some iron filings had been sprinkled into the tube, and many of these had stuck to the inside of the tube. With the same levels of water in the tube and the trough, the student recorded the volume of air in the tube (100 cm³).

After a few days he returned to the apparatus, equalised the water levels as before and took a second reading of the volume of air in the tube (85 cm³). He then worked out how much of the air had been replaced by water.

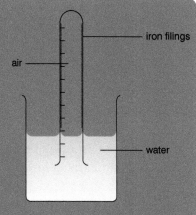

△ Fig. 10.4 Apparatus for experiment.

Observing, measuring and recording

1. Predict what changes you would expect to see in the iron filings that were put in the tube at the beginning of the experiment.

Interpreting observations and data

2. State what chemical reaction was taking place in the tube. Write a balanced equation for the reaction.

3. Explain what the results indicate about the composition of the air in the tube.

Evaluating methods

4. The results in this experiment are different to those obtained by other students. Suggest some possible reasons for this.

5. Explain why the student equalised the water levels in the trough and the tube before taking the reading of the volume of gas in the tube.

SCIENCE IN CONTEXT

THE MANUFACTURE OF OXYGEN AND NITROGEN

Oxygen, nitrogen and the noble gases can all be manufactured from the air. The industrial process has several stages.

1. The air is cooled to about −80 °C so that carbon dioxide and water vapour solidify and can be removed.

2. The air is cooled further, compressed and then allowed to expand quickly. This causes further cooling, and at about −200 °C the air becomes a liquid.

3. The liquid air is fractionally distilled. This involves using a large fractionating column which separates liquids with different boiling points. Oxygen's boiling point is −183 °C and nitrogen's is −196 °C: so they can be separated.

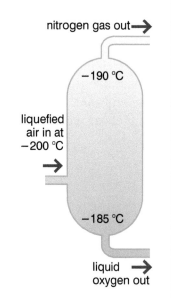

△ Fig. 10.5 Fractional distillation of liquid air.

Oxygen and nitrogen are stored in large metal cylinders under high pressure and remain as liquids until the pressure is released.

Oxygen is used in medicine to support patients with breathing difficulties. It is also mixed with hydrocarbons such as ethyne (acetylene) for use in cutting tools.

Nitrogen's uses depend on its very low reactivity. It is used in food packaging as a way of excluding air and stopping oxidation and decay. It is also used to reduce fire hazards, particularly in military aircraft fuel systems.

Challenge Question: Explain why oxygen is a liquid at $-185\,^{\circ}C$ and nitrogen is a gas at $-190\,^{\circ}C$.

QUESTIONS

1. Explain the term *anhydrous*.

2. Suggest what colour change you would observe if water is added to anhydrous cobalt(II) chloride.

3. Explain how you could test for the purity of a sample of water.

4. In the purification of water there are two important stages:

   a) In the first stage, the water is filtered twice. Describe what is used as the filter in each case.

   b) In the second stage, bacteria are killed. Describe which chemical is used to kill the bacteria.

5. This question is about the composition of the air.

   a) Describe the percentage of nitrogen in clean air.

   b) Describe the percentage of carbon dioxide in clean air.

Pollutants in the air

Pollutants in the air come from a variety of sources. Some come from burning waste and some from power stations burning coal or gas. Industry produces pollutants as well.

The most common pollutants in the air are:

- Carbon dioxide from the **complete combustion** of carbon-containing fuels. The increased levels of carbon dioxide result in **global warming** and **climate change**.
- Carbon monoxide and **particulates** from the **incomplete combustion** of carbon-containing fuels, for example, hydrocarbons (petrol/coal/gas/diesel). Carbon monoxide is a highly toxic and dangerous gas. Particulates increase the risk of respiratory problems and cancer.
- Methane from the decomposition of vegetation and waste gases from digestion in animals. Higher levels of methane lead to increasing global warming and climate change.
- Oxides of nitrogen from car engines. They contribute to the formation of **acid rain** and respiratory problems.
- Sulfur dioxide from the combustion of **fossil fuels** that contain sulfur compounds, for example, petrol and coal. Sulfur dioxide is a major source of acid rain.

Δ Fig. 10.6 Cycling is encouraged in Amsterdam to reduce air pollution.

How is the atmosphere changing and why?

Climate change is a huge issue facing all parts of the world.

SUPPLEMENT

The greenhouse effect

Carbon dioxide, methane and CFCs (chlorofluorocarbons) are known as **greenhouse gases**. The levels of these gases in the atmosphere are increasing due to the burning of fossil fuels, pollution from farm animals and the use of CFCs in aerosols and refrigerators.

Short-wave radiation from the Sun warms the ground, and the warm Earth emits thermal energy as long-wave radiation. Much of this radiation is stopped from escaping from the Earth by absorption and reflection. This reduction of thermal energy loss to space is known as the **greenhouse effect**.

The greenhouse effect is responsible for keeping the Earth warmer than it would otherwise be. This is known as global warming. If global warming continues the Earth's climate will change, polar ice will continue to melt and sea levels will continue to rise resulting in serious flooding of low-lying areas – some of them highly populated.

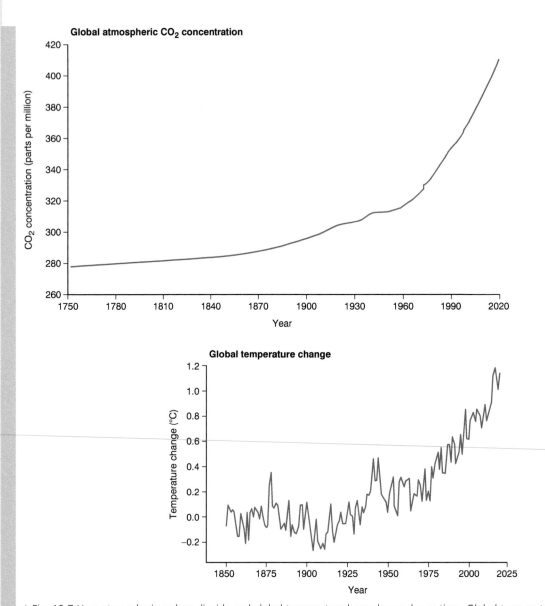

△ Fig. 10.7 How atmospheric carbon dioxide and global temperature have changed over time. Global temperature change is relative to temperatures in the pre-industrial period, 1850–1900.

Acid rain

Burning fossil fuels gives off many gases, including sulfur dioxide and various nitrogen oxides:

sulfur + oxygen → sulfur dioxide

$S(s) + O_2(g) → SO_2(g)$

Sulfur dioxide combines with water and oxygen to form sulfuric acid. Nitrogen oxide combines with water to form nitric acid. These substances can make the rain acidic (called acid rain).

sulfur dioxide + oxygen + water → sulfuric acid

$2SO_2(g) + O_2(g) + 2H_2O(l) → 2H_2SO_4(aq)$

Buildings are damaged by acid rain, particularly those made of limestone and marble – both are forms of calcium carbonate, $CaCO_3$. Metal structures are also attacked by sulfuric acid.

Acid rain harms plants that take in the acidic water and harms the animals that live in the affected rivers and lakes. Acid rain also washes ions such as calcium and magnesium out of the soil, depleting the minerals available to plants. It also washes aluminium ions out of the soil and into rivers and lakes, where they poison fish.

Nitrogen monoxide can be oxidised to a brown gas called nitrogen dioxide, NO_2. Nitrogen dioxide in the presence of water can form nitric acid. Under certain atmospheric conditions this can build up as a brown haze in large cities. Haze like this contain small chemical particles called particulates. **Particulates** and nitrogen dioxide are dangerous because they can cause respiratory diseases such as bronchitis.

Reducing emissions of the gases that cause acid rain is expensive. Part of the problem is that the acid rain usually falls a long way from the places where the gases were released.

Reducing the effects of climate change and acid rain

Steps are being taken to reduce the effects of environmental issues. These are summarised in the table below.

Environmental issues	Action being taken
Climate change	• increasing the use of hydrogen as a fuel • increasing the production of renewable energy using wind and solar power • planting trees to reduce the level of carbon dioxide by photosynthesis • reducing the use of fossil fuels • reduction in livestock farming to reduce methane production

SUPPLEMENT

Environmental issues	Action being taken
Acid rain	• increasing the use of catalytic converters in vehicles • reducing emissions of sulfur dioxide by using low-sulfur fuels and flue-gas desulfurisation with calcium oxide

△ Table 10.1 Reducing the effects of climate change and acid rain.

Under normal conditions nitrogen is a very unreactive gas. However, in a petrol engine temperatures of $1000\,°C$ can be reached, and in these conditions nitrogen reacts with oxygen to make oxides of nitrogen. Nitrogen oxides are often represented by NO_x, but they are mainly nitrogen monoxide, NO. A catalytic converter in the car's exhaust system can convert the nitrogen oxides back to nitrogen:

$$2NO(g) + 2CO(g) \rightarrow N_2(g) + 2CO_2(g)$$

Rhodium is frequently used as a catalyst in catalytic convertors.

SCIENCE IN CONTEXT SOME INTERESTING FACTS ABOUT METHANE

1. Methane makes up about 97% of natural gas.

2. It is formed by the decay of plant matter where there is no oxygen (anaerobic decay).

3. Biogas contains 40–70% methane. Biodigesters convert organic wastes into a nutrient-rich liquid fertiliser and biogas, a **renewable** source of electrical and thermal energy. These are widely used in countries such as India, Nepal and Vietnam. Biodigesters can help families by providing a cheap source of fuel, reducing environmental pollution from the run-off from animal pens, and reducing diseases caused by the use of untreated manure as fertiliser.

△ Fig. 10.8 A commercial biodigester.

4. Methane is one of the greenhouse gases responsible for global warming. It has almost 25 times the effect of the same volume of carbon dioxide.

5. Ruminant animals such as cows and sheep produce methane. It has been estimated that a cow can produce as much as 200 litres of methane per day.

Challenge Question: It seems highly likely that cows are one of the causes of global warming, Discuss the issues this raises for the world.

QUESTIONS

1. Carbon monoxide is a common pollutant in the air. Give a major source of the carbon monoxide.

2. **a)** An increase in methane emissions is one of the causes of global warming. Give two sources of methane.

   b) Give the name of another gas that is causing global warming.

3. **a)** Give the name of two gases that are responsible for causing acid rain.

   b) For each gas you named in part a) give the name of the acid it forms.

   c) Suggest two problems caused by acid rain in the environment.

4. SUPPLEMENT Nitrogen monoxide (NO) is produced from nitrogen and oxygen in the high temperature of a car engine.

   a) Write a balanced equation to show how the nitrogen monoxide is formed.

   b) Nitrogen monoxide and carbon monoxide are converted in a catalytic converter. Write an equation for this reaction.

End of topic checklist

Key terms

acid rain, anhydrous, combustion, climate change, fossil fuel, global warming, particulates, renewable energy

SUPPLEMENT greenhouse effect, greenhouse gas

During your study of this topic you should have learned:

○ How to describe chemical tests for identifying the presence of water using anhydrous cobalt(II) chloride and anhydrous copper(II) sulfate.

○ That the melting and boiling points of a water sample can be used to test its purity.

○ That distilled water is used in practical chemistry rather than tap water as it contains fewer chemical impurities.

○ How to describe in outline the treatment of the domestic water supply in terms of sedimentation, use of carbon and chlorination.

○ That clean dry air is approximately 78% nitrogen and 21% oxygen with the remainder made up of a mixture of noble gases and carbon dioxide.

○ That common pollutants in the air are carbon dioxide (from complete combustion of carbon-containing fuels), carbon monoxide and particulates (from incomplete combustion of carbon-containing fuels), methane (from decomposition of vegetation and waste gases from animal digestion), sulfur dioxide (from combustion of sulfur-containing fossil fuels) and oxides of nitrogen (from car engines).

○ That higher levels of carbon dioxide and methane lead to increased global warming and climate change.

○ That carbon monoxide is a toxic gas and particulates increase the risk of respiratory problems and cancer.

○ That sulfur dioxide and oxides of nitrogen cause acid rain.

○ That the oxides of nitrogen also cause photochemical smog and respiratory problems.

○ That climate change can be reduced by planting trees, reducing livestock farming, decreasing use of fossil fuels and increasing use of hydrogen and renewable energy.

○ **SUPPLEMENT** That acid rain can be reduced by using catalytic converters, using low-sulfur fuels and flue-gas desulfurisation with calcium oxide.

○ **SUPPLEMENT** How the greenhouse gases carbon dioxide and methane absorb and reflect thermal energy and so reduce heat loss to space.

○ **SUPPLEMENT** How to explain the removal of oxides of nitrogen by catalytic converters.

1. Explain why distilled water is used in practical chemistry rather than tap water.

2. **a)** State what you could use to detect the presence of water.

   b) State what you would observe if water was present.

3. In the purification of domestic water supplies state what chemical can be used to kill microbes.

4. This question is about the composition of a sample of clean, dry air.

   a) State the proportion of oxygen.

   b) State the proportion of carbon dioxide.

5. The table shows some of the pollutants found in air. For each pollutant identify the source of the pollution and how it gets into the air. The first one has been done for you.

Pollutant	What is the source of the pollutant?	How does the pollutant get into the air?
Lead compounds	Lead additives in petrol	Burning petrol in a car
Carbon monoxide		
Sulfur dioxide		
Oxides of nitrogen		

6. **SUPPLEMENT** This question is about the greenhouse effect.

   a) Explain what the greenhouse effect is.

   b) Give the names of two greenhouse gases.

7. **SUPPLEMENT** Explain how oxides of nitrogen formed in car engines can be removed by catalytic converters.

Practice questions for Sections 8, 9 and 10

Example answer

Note: Practice questions, sample answers and comments have been written by the authors. The marks awarded for these questions indicate the level of detail required in the answers. In examinations, the way marks are awarded may be different. References to assessment and/or assessment preparation are the publisher's interpretation of the syllabus requirements and may not fully reflect the approach of Cambridge Assessment International Education.

Question 1

Lithium (Li), sodium (Na) and potassium (K) are in Group I of the Periodic Table.

a) These elements have similar chemical properties.

Explain why, using ideas about electronic configurations.

All the elements have one electron in their outer shell. ✓ ① (1)

b) Lithium reacts with water to form a solution of lithium hydroxide and a colourless gas. During this reaction the temperature of the water increases.

i) What is the name of the colourless gas produced?

hydrogen ✓ ① (1)

ii) Why does the temperature of the water increase?

The reaction between lithium and water is rapid. ✗ (1)

COMMENTS

a) It would have been correct to state that all the elements have the same number of electrons in their outer shell. This answer goes further and correctly states that they all have one electron in the outer shell.

b) **i)** The correct answer has been given.

ii) This answer lacks precision. The answer would have been improved by explaining that the reaction produced heat or that the reaction was exothermic.

iii) Apart from stating that the reaction would be more rapid than that with lithium, observations have not been given. The products have been correctly named but these are not what you would *observe*. The answer would have been improved by mentioning: fizzing/effervescence/bubbles (of gas), the sodium floats/moves around on the water, forms a ball/disappears.

iv) The correct answer and explanation have been given. Caesium (Cs) is another Group I metal.

iii) Describe what you would observe if a small piece of sodium is added to water.

The sodium forms sodium hydroxide and hydrogen gas is given off. ✗

The reaction would be more rapid than the lithium reaction. ✓ ① (2)

iv) Is caesium more or less reactive than lithium? Give a reason for your answer.

More reactive, because reactivity in Group I increases down the group. ✓ ① (1)

(Total 6 marks)

Question 2

State which of the following gases causes increased global warming:

A Nitrogen

B Carbon monoxide

C Methane (1)

D Sulfur dioxide (Total 1 mark)

Question 3

Use the Periodic Table to help you answer this question.

a) State the symbol of the element with proton number 14. (1)

b) State the symbol of the element that has a relative atomic mass of 32. (1)

c) State the number of the group that contains the alkali metals. (1)

d) State which group contains elements whose atoms form ions with a 2+ charge. (1)

e) State which group contains elements whose atoms form ions with a 1– charge. (1)

(Total 5 marks)

Question 4

The reactivity of metals can be compared by comparing their reactions with dilute sulfuric acid. Pieces of zinc, iron and magnesium of identical size are added to separate test-tubes containing this acid.

a) Predict what order of reactivity would you expect. Put the most reactive metal first. (1)

b) Write a word equation for the reaction between magnesium and dilute sulfuric acid. (1)

c) Give the name of a metal that does not react with dilute sulfuric acid. (1)

d) State what other reaction could be used to compare the reactivity of metals. (1)

(Total 4 marks)

Question 5 SUPPLEMENT

Iron is extracted from iron ore in a blast furnace. Label the diagram of the blast furnace by selecting from the following possibilities.

bauxite	hematite	sodium hydroxide	cryolite
molten iron	sand	slag	

a)

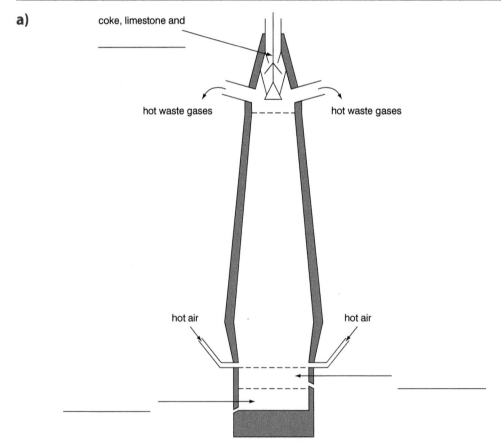

coke, limestone and
_____

hot waste gases   hot waste gases

hot air   hot air

(3)

b) Coke (carbon) burns in the oxygen in the furnace.

   i) Write a balanced equation for carbon burning in a plentiful supply of oxygen. (1)

   ii) The product formed in reaction **i)** above then reacts with more carbon to form a gas. Write a balanced equation for this reaction. (1)

c) Explain why limestone is added to the furnace. (2)

d) Iron is produced in the furnace by the reduction of iron(III) oxide. An equation for the reaction is:

$$Fe_2O_3(s) + 3CO(g) \rightarrow 2Fe(s) + 3CO_2(g)$$

Explain how you know that the iron(III) oxide has been reduced. (1)

e) Aluminium is another important metal.

   i) Unlike iron, aluminium cannot be obtained from aluminium ore by heating it with carbon. Explain why. (2)

   ii) State one large scale use of aluminium and a property of aluminium that makes it suitable for this use. (2)

(Total 12 marks)

Question 6

The table shows the composition of the mixture of gases in a typical car exhaust fumes.

Gas	% of the gas in the exhaust fumes
Carbon dioxide	9
Carbon monoxide	5
Oxygen	4
Hydrogen	2
Hydrocarbons	0.2
Nitrogen oxides	0.2
Sulfur dioxide	less than 0.003
Gas X	79.6

a) State the name of the gas X. (1)

b) The carbon dioxide comes from the burning of hydrocarbons, such as octane, in the petrol.

   i) Copy and complete the word equation for the complete combustion of octane.

   octane + _____ → carbon dioxide + _____ (2)

   ii) State which two elements are present in hydrocarbons. (1)

c) Suggest a reason for the presence of carbon monoxide in exhaust fumes. (1)

d) Nitrogen oxides are present in small quantities in exhaust fumes.

    i) Copy and complete the following equation for the formation of nitrogen dioxide.

    $N_2(g)$ + _____ $O_2(g) \rightarrow$ _____ $NO_2(g)$ (1)

    ii) State one harmful effect of nitrogen dioxide on organisms. (1)

e) Sulfur dioxide is an atmospheric pollutant and is found in only small amounts in car exhaust fumes.

    i) Give the main source of sulfur dioxide pollution of the atmosphere. (1)

    ii) Sulfur dioxide is oxidised in the air to sulfur trioxide. The sulfur trioxide may dissolve in rainwater to form a dilute solution of sulfuric acid, H_2SO_4. State the meaning of the term 'oxidation'. (1)

    iii) Sulfuric acid reacts with metals such as iron. Copy and complete the following word equation for the reaction of sulfuric acid with iron.

    sulfuric acid + iron $\rightarrow$ _____ + _____ (2)

    iv) Describe what effect acid rain has on buildings made of stone containing calcium carbonate. (1)

(Total 12 marks)

Question 7

Look at the list of five elements below:

argon, bromine, chlorine, iodine, potassium

a) Put these five elements in order of increasing proton number. (1)

b) Put these five elements in order of increasing relative atomic mass. (1)

c) The orders of proton number and relative atomic mass for these five elements are different. State which one of the following is the most likely explanation for this:

    A The proton number of a particular element may vary.

    B The presence of neutrons

    C The atoms easily gain or lose electrons.

    D The number of protons must always equal the number of neutrons. (1)

d) State which of the five elements in the list are in the same group of the Periodic Table. (1)

e) **i)** From the list, choose one element that has one electron in its outer shell. (1)

    ii) From the list, choose one element that has a full outer shell of electrons. (1)

(Total 6 marks)

Question 8 SUPPLEMENT

Analyse the information about the elements in Group I of the Periodic Table.

Element	Boiling point (°C)	Density (g/cm³)	Radius of atom in the metal (nm)	Reactivity with water
Lithium	1342	0.53	0.157	
Sodium	883	0.97	0.191	Rapid
Potassium	760	0.86	0.235	Very rapid
Rubidium		1.53	0.250	Extremely rapid
Caesium	669	1.88		Explosive

a) Describe how the density of the Group I elements changes going down the group. (2)

b) Suggest a value for the boiling point of rubidium. (1)

c) Suggest a value for the radius of a caesium atom. (1)

d) Use the information in the table to compare how fast lithium and the other Group I metals react with water. (1)

e) State three properties shown by all metals. (3)

f) When sodium reacts with water, hydrogen is given off:

$$2Na(s) + 2H_2O(l) \rightarrow 2NaOH(aq) + H_2(g)$$

i) State the name of the other product formed in this reaction. (1)

ii) Describe a test for hydrogen. (2)

(Total 11 marks)

Question 9 SUPPLEMENT

Explain using diagrams how alloys can be harder or stronger than pure metals. (4)

(Total 4 marks)

Question 10 SUPPLEMENT

a) Identify the greenhouse gas that forms during the complete combustion of hydrocarbons. (1)

b) Give one other example of a greenhouse gas. (1)

c) Describe how greenhouse gases cause global warming. (1)

d) Explain one way in which levels of greenhouse gases in the atmosphere can be reduced. (1)

(Total 4 marks)

Organic chemistry focuses almost entirely on the chemistry of covalently bonded carbon molecules. As such there are strong links with covalent bonding in Section 2, Atoms, elements and compounds. As well as life processes, organic chemistry includes the chemistry of other types of compounds including plastics, petrochemicals, drugs and paint. An understanding of organic chemistry begins with knowledge of the structure of a carbon atom and how it can combine with other carbon atoms by forming covalent bonds. In this section you will study the chemistry of a few of the 'families' or 'homologous series' of organic compounds, including hydrocarbons, alcohols and polymers. This will provide a firm foundation for further study in chemistry and biology.

Starting points

1. Where is carbon in the Periodic Table of elements?

2. What is the electronic configuration of carbon? How are its electrons arranged?

3. How does carbon form covalent bonds? Show the bonding in methane (CH_4), the simplest of organic molecules.

4. What do you know already about methane?

SYLLABUS SECTIONS COVERED

11.1 Formulas and terminology

11.2 Naming organic compounds

11.3 Fuels

11.4 Alkanes

11.5 Alkenes

11.6 Alcohols

11.7 Polymers

11
Organic chemistry

Δ Oil rigs are used to extract hydrocarbons from the Earth.

Fuels

INTRODUCTION

The most common fuels used today are either fossil fuels or are made from fossil fuels. There are problems associated with using fossil fuels – burning them produces a number of polluting gases and releases carbon dioxide, a greenhouse gas. Nevertheless, fossil fuels are still a very important source of energy.

Δ Fig. 11.1 Crude oil contains a mixture of hydrocarbons.

KNOWLEDGE CHECK

✓ Carbon compounds are covalently bonded.
✓ Covalent bonds are formed when non-metallic elements share electrons.
✓ Chemical formulas show the number and type of atoms in a molecule.
✓ Burning fossil fuels produces carbon dioxide, higher levels of carbon dioxide are leading to climate change.
✓ Burning some fossil fuels can also produce pollutant gases such as sulfur dioxide and nitrogen oxides.
✓ SUPPLEMENT Using renewable energy sources instead of fossil fuels will help reduce the effects of climate change.

LEARNING OBJECTIVES

✓ Draw and interpret the displayed formula of a molecule to show all the atoms and all the bonds.
✓ Name the fossil fuels: coal, natural gas and petroleum.
✓ Name methane as the main constituent of natural gas.
✓ State that hydrocarbons are compounds that contain hydrogen and carbon only.
✓ State that petroleum is a mixture of hydrocarbons.
✓ Describe the separation of petroleum into useful fractions by fractional distillation.
✓ SUPPLEMENT Describe how the properties of fractions obtained from petroleum change from the bottom to the top of the fractionating column, limited to: decreasing chain length; lower boiling points.
✓ Name the uses of the fractions as:
  a) refinery gas fraction for gas used in heating and cooking
  b) gasoline / petrol fraction for fuel used in cars
  c) naphtha fraction as a chemical feedstock
  d) diesel oil / gas oil for fuel used in diesel engines
  e) bitumen for making roads.

What are fossil fuels?

Petroleum (crude oil), natural gas (mainly methane) and coal are **fossil fuels**.

Crude oil was formed millions of years ago from the remains of animals and plants that were pressed together under layers of rock. It is usually found deep underground, trapped between layers of rock that it can't seep through (impermeable rock). Natural gas is often trapped in pockets above crude oil.

△ Fig. 11.2 Fractional distillation takes place in oil refineries, like this one in the Netherlands.

The supply of fossil fuels is limited – having taken millions of years to form, these fuels will eventually run out. They are called finite or **non-renewable** fuels. This makes them an extremely valuable resource that must be used efficiently.

Petroleum is a mixture of **hydrocarbons**, which are compounds that contain hydrogen and carbon only. These can be separated into useful **fractions** by the process of fractional distillation.

Fractional distillation

The chemicals in petroleum are separated into useful fractions by a process known as **fractional distillation**.

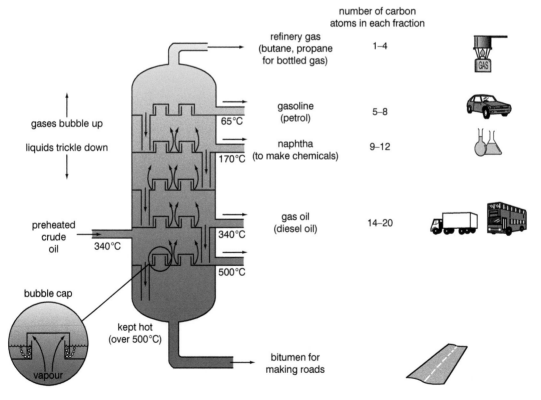

△ Fig. 11.3 A fractionating column separates crude oil into many useful fractions.

The crude oil is heated in a furnace and passed into the bottom of a fractionating column. It gives off a mixture of vapours that rise up the column, and the different fractions condense out at different heights. The fractions that come off near the top are light-coloured, runny liquids. Those removed near the bottom of the column are dark and sticky.

SUPPLEMENT

How does fractional distillation work?

The components of petroleum separate because they have different boiling points. A simple particle model explains why their boiling points differ. The different hydrocarbon molecules in petroleum are chemically bonded in similar ways with strong covalent bonds. However, they contain different numbers of carbon atoms.

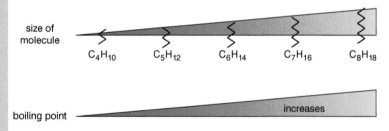

heptane

octane

△ Fig. 11.4 Octane has one more carbon atom and two more hydrogen atoms than heptane. Their formulas differ by CH_2.

The weak attractive forces between the molecules must be broken for the hydrocarbon to boil. The longer a hydrocarbon molecule is, the stronger the intermolecular forces between the molecules. The stronger these forces of attraction, the higher the boiling point because more energy is needed to overcome the forces.

size of molecule

C_4H_{10} C_5H_{12} C_6H_{14} C_7H_{16} C_8H_{18}

boiling point   increases

△ Fig. 11.5 How the properties of hydrocarbons change as chain length increases.

Other differences between the fractions include how easily they burn, how smoky their flames are.

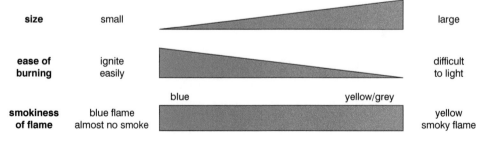

size	small		large
ease of burning	ignite easily		difficult to light
smokiness of flame	blue flame almost no smoke	blue ... yellow/grey	yellow smoky flame

△ Fig. 11.6 How different hydrocarbons burn.

1. Petroleum is a 'non-renewable' fuel. Explain what this means.

2. SUPPLEMENT One of the oil fractions obtained from the fractional distillation of crude oil is a liquid at room temperature and another is a gas at room temperature. Explain which fraction is more likely to have a small chain of carbon atoms.

3. SUPPLEMENT Another of the oil fractions obtained from the fractional distillation of petroleum burns with a very sooty yellow flame. Explain if this fraction is more likely to have a small chain of carbon atoms or a long chain.

SCIENCE IN CONTEXT

THE FOSSIL FUEL DILEMMA

Supplies of the non-renewable fossil fuels – oil, natural gas and coal – will eventually run out. However, it is not easy to estimate exactly when they will run out. Many different factors need to be considered, including how much of each deposit is left in the Earth, how fast we are using each fossil fuel at the moment, whether or not countries that have supplies will sell to those that don't, and how this is likely to change in the future. If we continue to switch to alternative fuel sources that are renewable, the reserves that we currently have will last longer.

Current estimates suggest that crude oil (petroleum) will run out before 2070. The estimate for natural gas is similar, with 2060 a possible date.

The situation with coal is very different. Most coal deposits have not yet been tapped, and the decline of the coal mining industry in countries such as the UK means that many coal seams are lying undisturbed. If we carry on using coal at the same rate as we do today, there could be enough to last well over a thousand years. However, as other fossil fuels run out, particularly oil, the use of coal may increase, reducing that timespan considerably but also contributing to global warming.

So, should we increase our efforts to develop renewable forms of energy such as wind and solar energy; should we put greater emphasis on nuclear power; or should we plan to make much greater use

△ Fig. 11.7 A coal-fired power station.

of coal? Solving this dilemma is likely to depend as much on political decisions as scientific ones.

Challenge Question: Discuss the options available for using fossil fuels in the future.

End of topic checklist

Key terms

crude oil, fraction, fractional distillation, fossil fuel, hydrocarbon, non-renewable, organic chemistry, petroleum

During your study of this topic you should have learned:

◯ That coal, natural gas and petroleum (crude oil) are fossil fuels.

◯ That methane is the main constituent of natural gas.

◯ That hydrocarbons are compounds that contain hydrogen and carbon only and petroleum is a mixture of hydrocarbons.

◯ How to describe the separation of petroleum into useful fractions by fractional distillation.

◯ **SUPPLEMENT** That the properties of fractions obtained from petroleum, including chain length, and boiling point, change from the bottom to the top of the fractionating column.

◯ That the uses of the fractions obtained from petroleum include:
  * refinery gas for gas in heating and cooking
  * gasoline/petrol for fuel in cars
  * naphtha for making chemicals
  * diesel/gas oil for fuel in diesel engines
  * bitumen for making roads.

End of topic questions

1. State the main component in natural gas:

    A butane   **B** propane   **C** methane   **D** octane

2. The diagram shows a column used to separate the components present in petroleum.

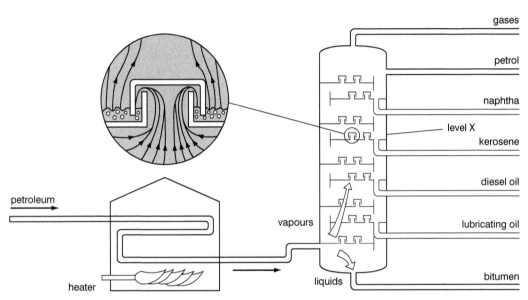

a) Give the name of the process used to separate petroleum into fractions.

b) **SUPPLEMENT** Describe what happens to the boiling point of the mixture as it goes up the column.

c) **SUPPLEMENT** The mixture of vapours arrives at level X. Explain what now happens to the various parts of the mixture.

3. This question is about octane, a component of petrol. Octane has a molecular formula of C_8H_{18}.

a) State the products formed when octane burns in a plentiful supply of air.

b) Write a word equation, including state symbols, for the reaction when octane reacts with a plentiful supply of oxygen.

c) Complete and balance the symbol equation, including state symbols, for the reaction when octane burns in a plentiful supply of air.

$$C_8H_{18}(l) + O_2(g) \rightarrow$$

Alkanes

INTRODUCTION

Alkanes are the simplest family or **homologous series** of **organic molecules**. The first alkane, methane, is the major component of natural gas, a common fossil fuel. Other alkanes are obtained from petroleum and are widely used as fuels.

Δ Fig. 11.8 Natural gas is being extracted from beneath the ocean floor.

KNOWLEDGE CHECK

✓ Covalent bonds are formed when non-metals bond together.
✓ Combustion involves burning in oxygen or air.

LEARNING OBJECTIVES

✓ **SUPPLEMENT** State that a homologous series is a family of similar compounds with similar chemical properties.
✓ State that a saturated compound has molecules in which all carbon–carbon bonds are single bonds.
✓ **SUPPLEMENT** Describe the general characteristics of a homologous series as: having the same general formula (recall of specific general formulas is not required); displaying a trend in physical properties; sharing similar chemical properties.
✓ Name and draw the displayed formulas of: (a) methane and ethane (the remainder of this Learning objective is covered in the topics on Alkenes and Alcohols).
✓ State the type of compound present, given a chemical name ending in *-ane*, *-ene* and *-ol*, *or* from a molecular formula or displayed formula (part of this Learning objective is covered in the topics on Alkenes and Alcohols).
✓ **SUPPLEMENT** Name and draw the structural and displayed formulas of unbranched: (a) alkanes containing up to four carbon atoms per molecule (the remainder of this Learning objective is covered in the topic on Alkenes).
✓ State that the bonding in alkanes is single covalent and that alkanes are saturated hydrocarbons.
✓ Describe the properties of alkanes as being generally unreactive, except in terms of combustion and substitution by chlorine.

WHAT ARE ALKANES?

Alkanes are **hydrocarbons**, which are molecules that contain only carbon and hydrogen atoms. They are made up of carbon atoms linked together by only single covalent bonds and are known as **saturated hydrocarbons**.

Many alkanes are obtained from crude oil by fractional distillation. The smallest alkanes are used extensively as fuels. Apart from combustion however, they are remarkably unreactive.

Alkane	Molecular formula	Displayed formula	Boiling point (°C)	State at room temperature and pressure
Methane	CH_4	H | H—C—H | H	−162	Gas
Ethane	C_2H_6	H H | | H—C—C—H | | H H	−89	Gas

Alkane	Molecular formula	Displayed formula	Boiling point (°C)	State at room temperature and pressure
Propane	C_3H_8	H H H | | | H—C—C—C—H | | | H H H	−42	Gas
Butane	C_4H_{10}	H H H H | | | | H—C—C—C—C—H | | | | H H H H	0	Gas

△ Table 11.1 Alkanes and their molecular structure.

Table 11.1 shows the names, molecular formulas and displayed formulas of some alkanes. Each line in a **displayed formula** represents a bond between atoms. Notice that the names of all alkanes end in -ane.

The **structural formula** is a formula that shows how the atoms in a molecule are arranged, without showing the bonds. To do this, look at the carbon atoms in the displayed formula, and the atoms attached to them. For example a molecule of ethane can be represented as two different sections each with 1 carbon atom and 3 hydrogen atoms. Similarly a molecule of propane can be represented as three different sections: two with 1 carbon atom and 3 hydrogen atoms and one with 1 carbon atom and 2 hydrogen atoms. This means we can write the structural formula of ethane as CH_3CH_3 and the structural formula of propane as $CH_3CH_2CH_3$.

Homologous series

Alkanes form a **homologous series**. Members of a homologous series have similar chemical properties.

The general characteristics of a homologous series include:

- Having the same general formula. For alkanes this is C_nH_{2n+2} (n equals the number of carbon atoms).
- Having similar chemical properties.
- Showing a gradual change in physical properties, such as melting point and boiling point (see Table 11.1).

△ Fig. 11.9 Formula 1 cars use specially blended mixtures of alkane hydrocarbons.

The properties of alkanes

The properties of alkanes are given in Table 11.2.

	Properties of alkanes
Description	Saturated (no double C=C bond)
Combustion	Burn in oxygen to form carbon monoxide (CO_2) and H_2O (CO if limited supply of oxygen)
Reactivity	Low
Chemical test	None
Uses	Fuels

△ Table 11.2 Properties of alkanes.

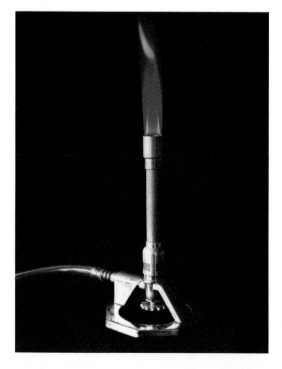

◁ Fig. 11.10 Methane is burning in the oxygen in the air to form carbon dioxide and water.

1. Alkanes are saturated hydrocarbons.

   a) Explain the term *saturated*.

   b) Explain the term *hydrocarbon*.

2. SUPPLEMENT a) Suggest the chemical formula for the alkane with 15 carbon atoms.

   b) Suggest what products you would expect to be formed if this alkane were burned in a plentiful supply of oxygen.

Combustion of alkanes

In a plentiful supply of air, alkanes burn to form carbon dioxide and water. A blue flame, as produced by a Bunsen burner, indicates **complete combustion**:

methane + oxygen → carbon dioxide + water

$$CH_4(g) + 2O_2(g) \rightarrow CO_2(g) + 2H_2O(l)$$

When the oxygen supply is limited, as when a Bunsen burner burns with a yellow flame, **incomplete combustion** occurs:

methane + oxygen → carbon + water

$$CH_4(g) + O_2(g) \rightarrow C(s) + 2H_2O(l)$$

The incomplete combustion of hydrocarbons such as methane can be very dangerous. It can produce carbon monoxide, which is extremely poisonous.

methane + oxygen → carbon monoxide + water

$$2CH_4(g) + 3O_2(g) \rightarrow 2CO(g) + 4H_2O(l)$$

Carbon monoxide is difficult to detect without special equipment, because it has no colour or smell. Gas and oil heaters or boilers must be serviced regularly. This is to ensure that jets do not become blocked and limit the air supply, or that exhaust flues don't become blocked and allow small quantities of carbon monoxide to enter the room. The flame in such a boiler or heater should always be blue in colour.

Δ Fig. 11.11 The gas in this cooker is burning completely.

In addition, alkanes will undergo substitution reactions with chlorine.

1. Explain how you can tell if a fuel is burning without enough oxygen.

2. Give the names of two products that can form when methane burns in an insufficient supply of oxygen.

3. Suggest some alternative ways of generating electricity that do not involve burning fossil fuels.

End of topic checklist

Key terms

alkane, combustion, complete combustion, displayed formula, hydrocarbon, incomplete combustion, organic molecules, saturated hydrocarbon

SUPPLEMENT homologous series, structural formula

During your study of this topic you should have learned:

○ **SUPPLEMENT** That a homologous series is a family of similar compounds with similar chemical properties.

○ That a saturated compound has molecules in which all carbon–carbon bonds are single bonds.

○ How to describe names of methane and ethane and how to draw their displayed formulas.

○ That a chemical with a name ending in '-ane' is an alkane.

○ That the bonding in alkanes is single covalent and alkanes are saturated hydrocarbons.

○ That alkanes are generally unreactive apart from combustion.

○ **SUPPLEMENT** That general characteristics of a homologous series include having the same and general formula, sharing similar chemical properties and showing a trend in physical properties.

○ **SUPPLEMENT** How to name and draw the structural and displayed formulas of unbranched alkanes containing up to four carbon atoms per molecule.

End of topic questions

1. **SUPPLEMENT** The alkane propane has the formula C_3H_8. Suggest which of the following is the general formula for the alkanes:

   A C_nH_{2n}

   B C_nH_{2n+2}

   C C_nH_{2n+4}

   D C_nH_n

2. **SUPPLEMENT** Define the term *homologous series*.

3. Ethane burns in oxygen.

   a) State the products formed when ethane burns in excess oxygen.

   b) Describe what colour flame would indicate that the ethane was burning in excess oxygen.

   c) Complete the balanced equation for the reaction:

   $$C_2H_6 + O_2 \rightarrow$$

   d) State the names of two additional products that could be formed if the oxygen supply was limited.

   e) Describe what colour flame would indicate that the ethane was burning in a limited supply of oxygen.

Alkenes

INTRODUCTION

Alkenes are hydrocarbons and burn in air in the same way that alkanes do. However, alkenes are much more reactive due to the C=C double bond they contain. This makes them very useful starting materials for a number of important industrial processes, including the manufacture of synthetic polymers.

Δ Fig. 11.12 The plastic objects above are made from alkenes.

KNOWLEDGE CHECK

✓ An unsaturated compound has molecules containing a double or triple covalent bond.
✓ **SUPPLEMENT** A homologous series is a family of similar compounds with similar chemical properties.
✓ A displayed formula shows all the atoms and bonds in the molecule.

LEARNING OBJECTIVES

✓ State that an unsaturated compound has molecules in which one or more carbon–carbon bonds are not single bonds.
✓ Name and draw the displayed formula of: (b) ethene (the remainder of this Learning objective is covered in the topics on Alkanes and Alcohols).
✓ State the type of compound present, given a chemical name ending in *-ane*, *-ene* and *-ol*, *or* from a molecular formula or displayed formula (part of this Learning objective is covered in the topics on Alkanes and Alcohols).
✓ **SUPPLEMENT** Name and draw the structural and displayed formulas of unbranched: (b) alkenes, including but-1-ene and but-2-ene (**not** cis/trans), containing up to four carbon atoms per molecule (the remainder of this Learning objective is covered in the topic on Alkanes).
✓ State that the bonding in alkenes includes a double carbon–carbon covalent bond and that alkenes are unsaturated hydrocarbons.
✓ **SUPPLEMENT** Describe the manufacture of alkenes and hydrogen by the cracking of larger alkane molecules using a high temperature and a catalyst.
✓ Describe the test to distinguish between saturated and unsaturated hydrocarbons by their reaction with aqueous bromine.
✓ **SUPPLEMENT** Describe the properties of alkenes in terms of addition reactions with bromine or aqueous bromine; hydrogen in the presence of a nickel catalyst; steam in the presence of an acid catalyst, and draw the structural or displayed formulas of the products.

WHAT ARE ALKENES?

Alkenes contain one or more carbon-to-carbon double bonds. Hydrocarbons with at least one double bond are known as **unsaturated hydrocarbons**. Alkenes burn well and are reactive in other ways also. Their reactivity is due to the carbon-to-carbon double bond.

SUPPLEMENT

Alkenes are another homologous series, so they have similar chemical properties and physical properties that change gradually from one member to the next.

Table 11.3 shows the names, molecular formulas and displayed formulas of some alkenes. Notice that = represents a double bond, and the names of all alkenes ends in –ene.

Alkene	Molecular formula	Displayed formula	Boiling point (°C)	State at room temperature and pressure
Ethene	C_2H_4	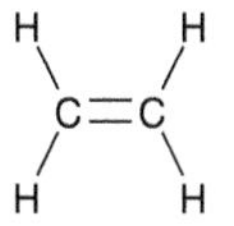	−104	Gas
Propene	C_3H_6	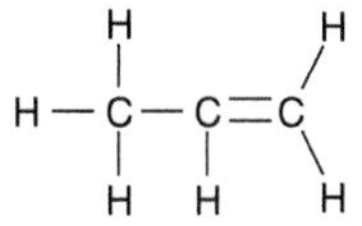	−48	Gas

△ Table 11.3 Structure and state of some alkenes.

SUPPLEMENT

The structural formula of ethene is $CH_2 = CH_2$ and the structural formula of propene is $CH_2 = CHCH_3$. Notice that for alkenes the structural formula also shows the location of the C=C double bond.

SUPPLEMENT

Cracking

Small hydrocarbon molecules are much more useful than larger molecules. Larger alkane molecules can be broken down into smaller ones by **catalytic cracking**. This requires a high temperature of between 600 to 700 °C and a catalyst of silica or alumina.

The composition of petroleum varies in different parts of the world. Table 11.4 shows the composition of a sample of petroleum from the Middle East after fractional distillation.

Fraction (in order of increasing boiling point)	Typical percentage produced by fractional distillation
Liquefied petroleum gases (LPG)	3
Gasoline	13
Naphtha	9
Kerosene	12
Diesel	14
Heavy oils and bitumen	49

Δ Table 11.4 Oil fractions.

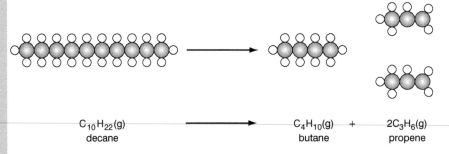

$C_{10}H_{22}(g)$ $C_4H_{10}(g)$ + $2C_3H_6(g)$
decane   butane   propene

Δ Fig. 11.13 The decane molecule ($C_{10}H_{22}$) is converted into the smaller molecules butane (C_4H_{10}) and propene (C_3H_6) in cracking.

QUESTIONS

1. Explain why cracking is needed in addition to the fractional distillation of crude oil.

2. Suggest the conditions needed for the cracking of oil fractions.

A simple test to distinguish alkenes from alkanes, or an **unsaturated hydrocarbon** from a **saturated** one, is to add bromine water, an orange solution, to the hydrocarbon. Alkanes do not react with bromine water, so the colour does not change. An alkene does react with the bromine, and the bromine water loses its colour.

Developing practical skills

A group of students set up an experiment to see if they could 'crack' some liquid paraffin. They soaked some mineral wool in liquid paraffin and assembled the apparatus as shown in Fig. 11.14. The students all wore eye protection until they had finished. They used a borosilicate boiling tube and heated it gently to begin with. They then heated the pottery pieces very strongly, occasionally letting the flame heat the mineral wool. Bubbles of gas started to collect in the test-tube. After a few minutes they had collected three test-tubes full of gas and so they stopped heating and immediately lifted the delivery tube out of the water.

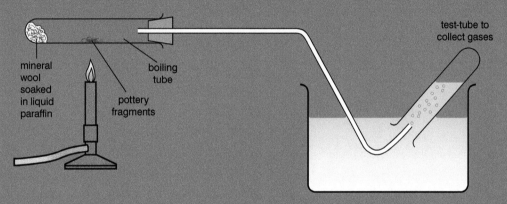

△ Fig. 11.14 Incorrectly set up apparatus for experiment.

Evaluating methods

1. The gas or gases produced in this reaction can be collected by displacement of water. State what property of gas(es) this demonstrates.

2. Explain why the delivery tube must be lifted out of the water as soon as heating stops.

Using and organising techniques, apparatus and materials

3. Describe the hazards involved in this experiment. Suggest what safety precautions would minimise them.

4. The first test-tube of gas collected did not burn, but the second one did. Explain this difference.

Interpreting observations and data

5. One of the students suggested that one of the two products was ethene (C_2H_4). Assuming that liquid paraffin has the formula $C_{14}H_{30}$, write an equation for the cracking of the liquid paraffin used in this experiment.

Addition reactions of alkenes

The reaction between an alkene and bromine water is known as an **addition reaction**:

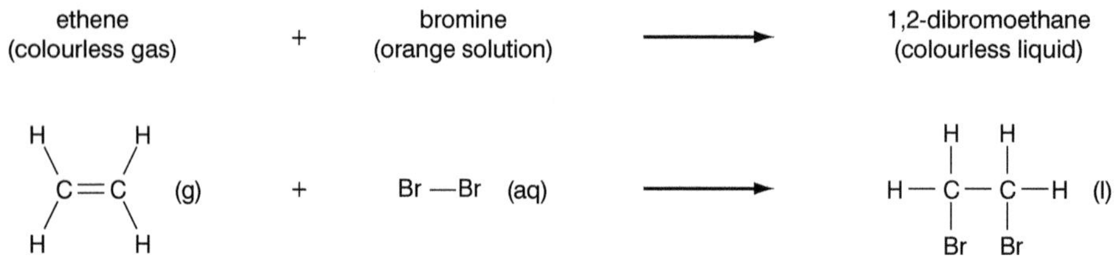

ethene   +   bromine   1,2-dibromoethane
(colourless gas)   (orange solution)   (colourless liquid)

△ Fig. 11.15 The reaction of ethene and bromine water.

This addition reaction occurs both with aqueous bromine (bromine water) or bromine. As shown in the equation, in an addition reaction only one product is formed.

Ethene undergoes other addition reactions. It also reacts with hydrogen in the presence of a nickel catalyst to form ethane:

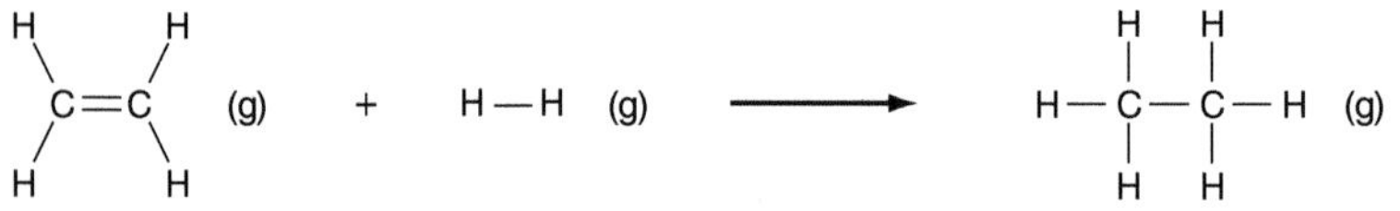

△ Fig. 11.16 Ethene reacts with hydrogen to form ethane.

$$C_2H_4(g) + H_2(g) \rightarrow C_2H_6(g)$$

Ethene also reacts with steam to form ethanol, using an acid catalyst such as phosphoric acid. This reaction is used in the manufacture of ethanol.

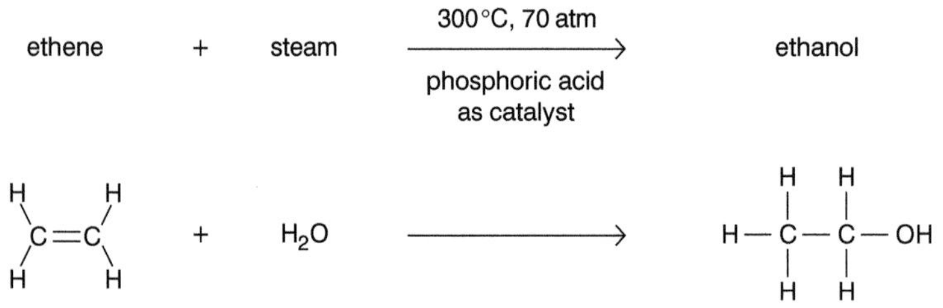

ethene   +   steam $\xrightarrow[\text{phosphoric acid as catalyst}]{300°C, 70 atm}$ ethanol

△ Fig. 11.17 Ethene reacts with steam to form ethanol.

The properties of alkenes

The properties of alkenes are given in Table 11.5.

	Properties of alkenes
Description	Unsaturated (contain a double C=C bond)
Combustion SUPPLEMENT	Burn in oxygen to form CO_2 and H_2O (CO if limited supply of oxygen)
Reactivity SUPPLEMENT	High (because of the double C=C bonds)
Chemical test	Turn bromine water from orange to colourless
Uses SUPPLEMENT	Making polymers such as polyethene

△ Table 11.5 Properties of alkenes.

Homologous series

Alkenes form another homologous series, so they have similar chemical properties and undergo addition reactions as described above, and physical properties that change gradually from one member to the next (Table 11.5).

Alkenes have the general formula C_nH_{2n} where n is the number of carbon atoms in the formula. It is not necessary to learn this formula as it is not on the syllabus, but notice that the formula only contains two types of atom. In Table 11.3 notice that ethene has two carbon atoms and four carbon atoms. If you put $n = 2$ in C_nH_{2n} you get the molecular formula C_2H_4.

Unbranched alkenes can have different displayed formulas.

Alkenes containing four carbon atoms can have two different arrangement of the atoms. This means they have the same molecular formula, C_4H_8, but different displayed formulas as shown below. These two different compounds are called but-1-ene and but-2-ene. Notice that they have the same number of carbon and hydrogen atoms, but the location of the C=C double bond is different.

Δ Fig. 11.18 But-1-ene and but-2-ene.

The structural formulas are $CH_3CH_2CH=CH_2$ (but-1-ene) and $CH_3CH=CHCH_3$ (but-2-ene).

QUESTIONS

1. Ethene is an unsaturated hydrocarbon. Explain the term *unsaturated*.

2. SUPPLEMENT Explain how ethene is manufactured.

3. SUPPLEMENT Explain why larger alkane molecules are cracked to produce alkene molecules.

4. SUPPLEMENT Write down the structural formula of propene.

SATURATED AND UNSATURATED FATS

We all need some **fat** in our diet because it helps the body to absorb certain nutrients. Fat is also a source of energy and provides essential fatty acids. However, it is best to keep the amount of fat we eat at sensible levels and to eat unsaturated fats rather than saturated fats whenever possible. A diet high in saturated fat can cause the level of cholesterol in the blood to build up over time. Raised cholesterol levels increase the risk of heart disease.

Foods high in saturated fat include:

- fatty cuts of meat
- meat products and pies
- butter
- cheese, especially hard cheese
- cream and ice cream
- biscuits and cakes.

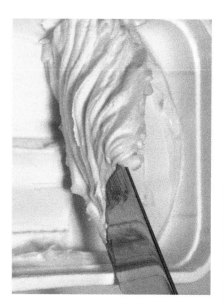

Δ Fig. 11.19 Margarine is made from vegetable oil, whose unsaturated molecules have been saturated with hydrogen.

Unsaturated fat is found in:

- oily fish such as salmon, tuna and mackerel
- avocados
- nuts and seeds
- sunflower and olive oils.

Challenge Question: Explain the main difference in the structure of saturated and unsaturated organic compounds.

End of topic checklist

Key terms

alkene, catalytic cracking, displayed formula, saturated hydrocarbon, unsaturated hydrocarbon

SUPPLEMENT addition reaction, homologous series, structural formula

During your study of this topic you should have learned:

○ How to name and draw the displayed formula of ethene.

○ That a chemical with a name ending in '-ene' is an alkene.

○ That the bonding in alkenes includes a double carbon–carbon covalent bond and that alkenes are unsaturated hydrocarbons.

○ SUPPLEMENT How to describe the manufacture of alkenes and hydrogen by the cracking of larger alkane molecules, using high temperature and a catalyst.

○ How to distinguish between saturated and unsaturated hydrocarbons from their molecular structures and by reaction with aqueous bromine.

○ SUPPLEMENT How to name and draw the structural and displayed formulas of unbranched alkenes, including but-1-ene and but-2-ene, containing up to four carbon atoms per molecule.

○ SUPPLEMENT How to describe the properties of alkenes in terms of addition reactions with bromine or bromine water, hydrogen (in the presence of a nickel catalyst) and steam (in the presence of an acid catalyst), and draw the structural or displayed formulas of the products.

End of topic questions

1. The reaction with aqueous bromine can be used to distinguish between an alkane and an alkene. State which of the following statements is correct:

   A An alkene will produce a dark brown solution with bromine water.

   B An alkane will decolourise bromine water.

   C Bromine water will change from orange to colourless in the presence of an alkene at room temperature.

   D An alkene will decolourise bromine water but only in the presence of UV light.

2. **SUPPLEMENT** Ethene burns in oxygen.

   a) State the names of the products formed when there is a plentiful supply of oxygen.

   b) i) Write a balanced equation for the burning of ethene in a plentiful supply of oxygen.

   ii) Describe what colour the flame would be.

   c) When ethene is burned in a limited supply of air, carbon and water are formed.

   i) Write a balanced equation for this reaction.

   ii) Describe what colour the flame would be.

3. **SUPPLEMENT a)** Draw displayed formulas for butane and butene.

   b) State which hydrocarbon is unsaturated.

   c) State which substance you could use to distinguish between butane and butene.

4. **SUPPLEMENT** The cracking of decane molecules is shown by the equation $C_{10}H_{22} \rightarrow Y + C_2H_4$

   a) Decane is a hydrocarbon. Define the term *hydrocarbon*.

   b) State what reaction conditions are needed for cracking.

   c) Write down the molecular formula for hydrocarbon Y.

   d) State which 'family' hydrocarbon Y belongs to.

   e) Explain why the cracking of petroleum fractions is so important.

5. Propene gas is bubbled through some bromine water.

a) Describe the colour change that would occur.

b) SUPPLEMENT Draw the displayed formulas of propene and bromine.

c) SUPPLEMENT Draw the displayed formula of the product of the reaction between propene and bromine water

Alcohols

INTRODUCTION

Ethanol, or common alcohol, belongs to a homologous series known as the alcohols. There are many more uses for ethanol than for other alcohols, and so its manufacture is very important.

△ Fig. 11.20 The ethanol being made here could be used as a solvent or as a fuel in cars.

KNOWLEDGE CHECK

✓ **SUPPLEMENT** Homologous series are families of similar compounds with similar chemical properties.
✓ Saturated compounds have molecules in which all the carbon–carbon bonds are single.
✓ Unsaturated compounds have molecules in which one or more carbon–carbon bonds are double or triple.

LEARNING OBJECTIVES

✓ Name and draw the displayed formula of: (c) ethanol (the remainder of this Learning objective is covered in the topics on Alkanes and Alkenes).
✓ State the type of compound present, given a chemical name ending in -ane, -ene and -ol, or from a molecular formula or displayed formula (part of this Learning objective is covered in the topics on Alkanes and Alcohols).
✓ Describe the complete combustion of ethanol.
✓ State the uses of ethanol as a solvent and as a fuel.

WHAT ARE ALCOHOLS?

Alcohols are hydrocarbon molecules that contain the OH group, of atoms attached to one of the carbon atoms.

SUPPLEMENT

Alcohols have the general formula $C_nH_{2n+1}OH$ and belong to the same homologous series, part of which is shown in Table 11.6.

Alcohol	Formula	Displayed formula	Boiling point (°C)
Methanol	CH_3OH	H | H—C—OH | H	65
Ethanol	C_2H_5OH	H H | | H—C—C—OH | | H H	78

Alcohol	Formula	Displayed formula	Boiling point (°C)
Propanol	C_3H_7OH	H―C―C―C―OH (propanol structure)	97
Butanol	C_4H_9OH	H―C―C―C―C―OH (butanol structure)	118

Δ Table 11.6 The first four alcohols.

Alcohols are another homologous series, so they have similar chemical properties and physical properties that change gradually from one member to the next (see Table 11.6).

Alcohols have the general formula $C_nH_{2n+1}OH$, where n represents the number of carbon atoms in the formula. It is not necessary to learn this formula as it is not on the syllabus, but notice that the formula ends in –OH.

Ethanol burns readily in air to form carbon dioxide and water. The equation for this complete combustion reaction is:

ethanol + oxygen → carbon dioxide + water

$$C_2H_5OH(l) + 3O_2(g) \rightarrow 2CO_2(g) + 3H_2O(l)$$

Ethanol – the most common alcohol

Ethanol, commonly just called 'alcohol', is the most widely used of the alcohol family. Two of its major uses are given in Table 11.7.

Use of ethanol	Reason
Solvent – such as in disinfectants and perfumes	The OH group allows it to dissolve in water, and it dissolves other organic compounds.
Fuel – such as for cars	It only releases CO_2 and H_2O into the environment, not other pollutant gases as from petrol. It is a renewable resource because it comes from plants, for example sugar beet and sugar cane.

Δ Table 11.7 Uses of ethanol.

◁ Fig. 11.21 Brazilians often use *alcool* as part of vehicle fuel (about 20%). It is made from the fermented and distilled juice of sugar cane.

QUESTIONS

1. Describe the formula of the alcohol ethanol.
2. Describe the advantages of using ethanol as a fuel in motor cars.
3. Ethanol is commonly used as a *solvent*. Explain this term.

Developing practical skills

A group of chemists needed to carry out the fermentation of sugar in the laboratory to make some ethanol for fuel. They set up the apparatus shown in Fig. 11.22. To prevent the oxidation of ethanol to vinegar, the apparatus was set up so that air could not enter the flask. At first there seemed to be nothing happening, but by the next morning the reaction had started. When the reaction had finished, the chemists used fractional distillation to obtain some pure alcohol.

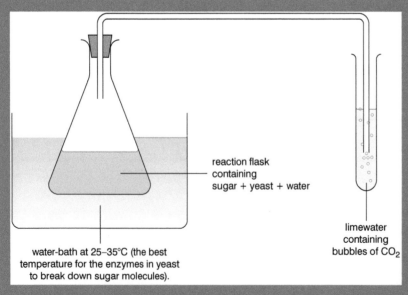

reaction flask containing sugar + yeast + water

limewater containing bubbles of CO_2

water-bath at 25–35°C (the best temperature for the enzymes in yeast to break down sugar molecules).

Fig. 11.22 Fermentation apparatus for making ethanol in the laboratory.

Evaluating methods

1. Describe how the chemists could have tried to maintain the temperature of the water-bath while the reaction was proceeding.

2. State the purpose of the limewater.

3. Describe what the chemists would have observed to indicate that a reaction had started.

4. Explain how the chemists would know when the reaction was complete.

5. In the fractional distillation process, state which liquid would be collected first – ethanol or water. Explain your answer.

ETHANOL AS A FUEL

The development of ethanol as a fuel for cars has a long history. In the USA, until 1908, some cars used ethanol as a fuel. Now most cars in the USA can run on blends of petrol/ethanol containing up to 10 percent ethanol.

△ Fig. 11.23 In the USA, the ethanol in gasohol is made from corn (maize).

Brazil and the USA are the World's top producers of ethanol. In the USA 97% of the gasoline sold has ethanol in it. In Brazil about one-fifth of all cars can use 100% ethanol fuel, known as E100; many other cars use petrol/ethanol blends. Brazil makes its ethanol mostly from fermented sugar obtained from sugar cane; in the USA, corn (maize) is used as the source of the sugar. Starting from sugar cane is much more efficient, and so ethanol is much cheaper to produce in Brazil than in the USA.

One problem with using ethanol as a fuel is that it readily absorbs water from the atmosphere. This causes some problems when transporting the fuel and makes it more expensive than petrol to transport.

A key question is, should food be used to make a fuel? Is it better to use fuel made from crude oil? Some people argue that food is needed for people to eat and should not be used to run cars. They say that using so much land for growing sugar cane and corn to make ethanol means that less land is available for growing food. This leads to food shortages and increases in the price of food. Additionally in Brazil, tropical rainforests have been cleared to grow sugar cane for use in ethanol production.

Challenge Question: Analyse the issues raised about producing ethanol from sugar cane.

End of topic checklist

Key terms

alcohol, displayed formula, solvent

SUPPLEMENT homologous series

During your study of this topic you should have learned:

○ How to name and draw the displayed formula of ethanol.

○ That a chemical with a name ending in '-ol' is an alcohol.

○ How to describe the complete combustion of ethanol to give carbon dioxide and water.

○ That ethanol is used as a solvent and as a fuel.

1. State which of the following equations correctly represents the combustion of ethanol in a plentiful supply of oxygen:

A $2C_2H_5OH + 2O_2 \rightarrow 2CO_2 + 3H_2O$

B $C_2H_5OH + 2O_2 \rightarrow 2CO + 3H_2O$

C $C_2H_5OH + O_2 \rightarrow 2C + 3H_2O$

D $C_2H_5OH + 3O_2 \rightarrow 2CO_2 + 3H_2O$

2. Which of the following substances has the following displayed formula:

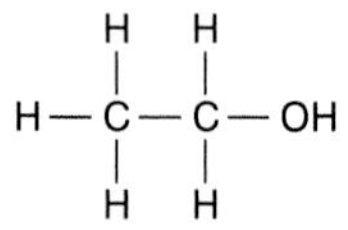

$$\begin{array}{ccc} & H & H \\ & | & | \\ H - & C - & C - OH \\ & | & | \\ & H & H \end{array}$$

A ethane

B ethene

C ethanol

D methanol?

3. State two main uses of ethanol.

Polymers

INTRODUCTION

A polymer is a very large molecule made up from smaller molecules joined together. The smaller molecules are called monomers. Polymers can have different numbers of monomers and have different links or bonds between the monomers. Synthetic polymers are those that are manufactured and some types are often referred to as plastics. Some types of synthetic polymers are used as fabrics and have names such as 'polyester' or 'polyamide'.

△ Fig. 11.24 Most tennis rackets have carbon fibre reinforced polymer frames.

KNOWLEDGE CHECK

✓ Alkenes are unsaturated hydrocarbons and contain carbon-to-carbon double bonds.
✓ Alkenes undergo addition reactions.
✓ Plastics cause damage to the environment.

LEARNING OBJECTIVES

✓ Define polymers as large molecules built up from many smaller molecules called monomers.
✓ Describe the formation of poly(ethene) as an example of addition polymerisation using ethene monomers.
✓ SUPPLEMENT Identify the repeat units in addition polymers and in condensation polymers.
✓ SUPPLEMENT Deduce the structure or the repeat unit of an addition polymer from a given alkene and vice versa.
✓ SUPPLEMENT Describe the differences between addition and condensation polymerisation.
✓ SUPPLEMENT Describe and draw the structure of nylon, a polyamide:

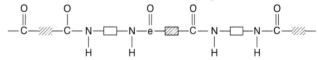

SYNTHETIC POLYMERS

There are two types of synthetic polymers: **addition polymers** and **condensation polymers**. Alkenes can be used to make **polymers**, which are very large molecules made up of many smaller molecules called **monomers**.

Addition polymers

Alkenes are able to react with themselves. They join together into long chains, like adding beads to a necklace. When the monomers add together like this, the material produced is called an addition polymer. Poly(ethene) or polythene is made this way.

By changing the atoms or groups of atoms attached to the carbon-to-carbon double bond, a range of different polymers can be made.

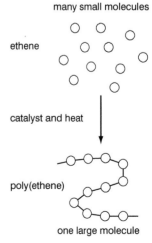

△ Fig. 11.25 Ethene molecules link together to produce a long polymer chain of poly(ethene).

The double bond within an alkene molecule breaks to form a single covalent bond to a carbon atom in an adjacent molecule. This process is repeated rapidly as the molecules link together.

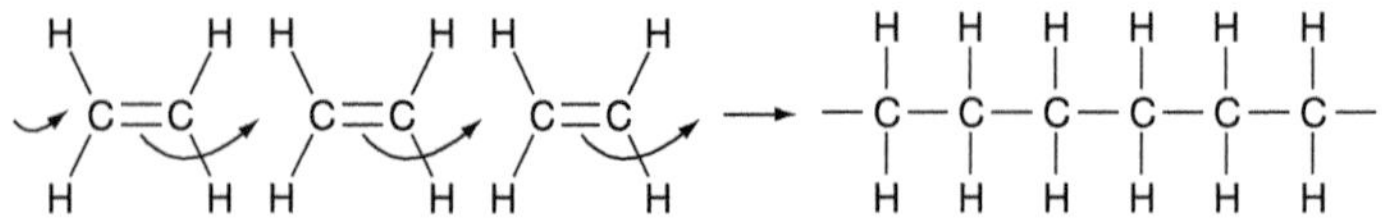

△ Fig. 11.26 Poly(ethene) from ethene.

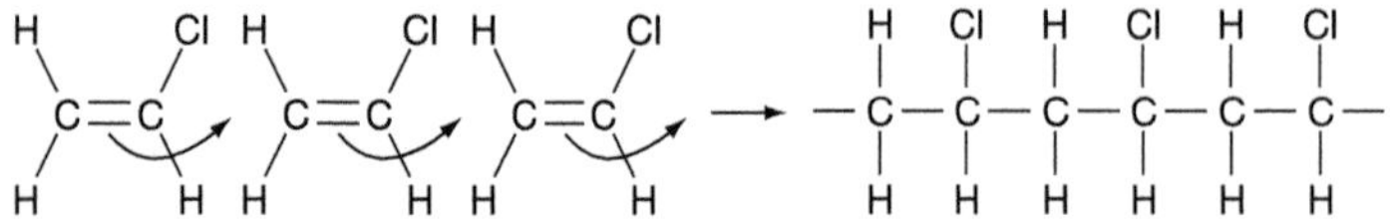

△ Fig. 11.27 Poly(chloroethene) from chloroethene.

SUPPLEMENT

Polymers are made from monomers. In some cases the polymer is made from just one monomer. In other polymers there are two or more different monomers.

The displayed formulas of the addition polymers in Table 11.8 show the repeat units and linkages in the polymers. You will see that the double bond in the alkene is replaced by each carbon atom in the double bond forming an additional single bond and so forming the linked structure of the polymer.

Name of monomer	Displayed formula of monomer	Name of polymer	Displayed formula of polymer 'showing the repeat unit (n is a large number)'	Uses of polymer
Ethene	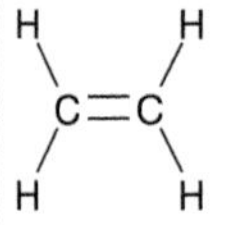	Poly(ethene)	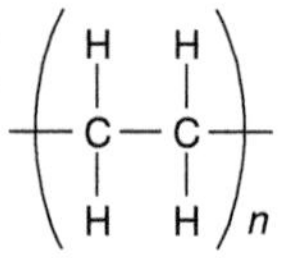	Buckets, bowls, plastic bags
Propene	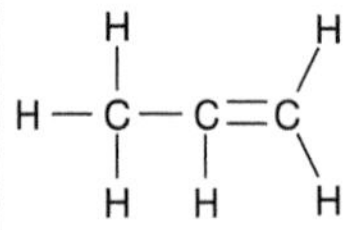	Poly(propene)	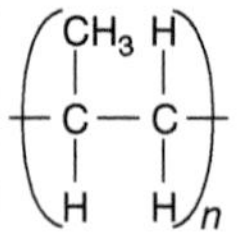	Packaging, ropes, carpets
Chloroethene (vinyl chloride)	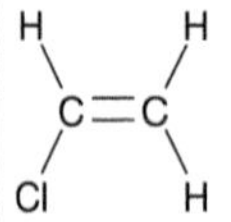	Poly-(chloroethene) (polyvinyl chloride)	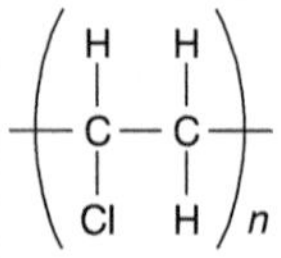	Plastic sheets, artificial leather
Phenylethene (styrene)	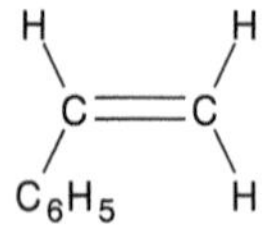	Poly-(phenylethene) (polystyrene)	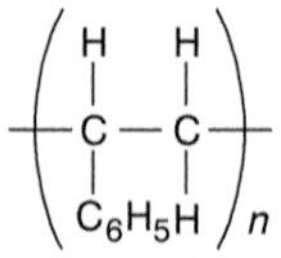	Yoghurt cartons, packaging
Tetrafluoroethene	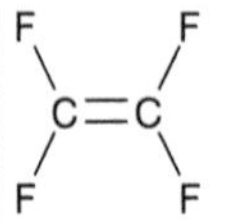	Poly-(tetrafluroethene) or PTFE	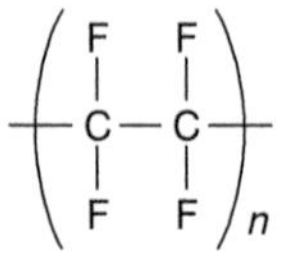	Non-stick coating in frying pans

△ Table 11.8 Monomers and their polymers.

△ Fig. 11.28 This pan is coated with PTFE non-stick plastic. How is PTFE made?

QUESTIONS

1. Explain in what ways a polymer is like a string of beads.

2. SUPPLEMENT The diagram shows the structural formula of chloroethene.

$$\underset{\underset{H}{|}}{\overset{\overset{H}{|}}{C}} = \underset{\underset{H}{|}}{\overset{\overset{Cl}{|}}{C}}$$

   a) Show how two molecules of chloroethene join together to form part of the polymer poly(chloroethene).

   b) Draw the structure of the repeat unit of the polymer.

3. SUPPLEMENT Suggest what type of polymer poly(chloroethene) is.

SUPPLEMENT

Condensation polymers

Addition polymers are made from one type of monomer, as in poly(ethene) for example. Polymers can also be made by joining together two different monomers so that they react together. When they react, they expel a small molecule. Because the molecule is usually water, the process is called condensation polymerisation and the products are condensation polymers.

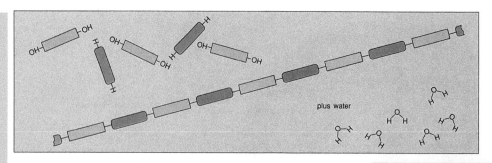

△ Fig. 11.29 Condensation polymerisation.

Nylon is a common synthetic condensation polymer.

Nylon is made from a molecule with two carboxylic acid groups (COOH) and another molecule with two amine groups (NH_2). Nylon is a **polyamide**.

They react to form an **amide** link. The amide link is formed when a carboxylic acid group (COOH) reacts with an amine group (NH_2) and water is formed along with the polymer chain.

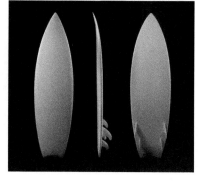

△ Fig. 11.30 Surfboards are made from a condensation polymer called polyurethane.

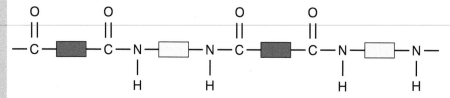

△ Fig. 11.31 Nylon is a polyamide.

Condensation polymerisation is used for making synthetic fibres such as nylon for clothes, ropes and other uses. When PET is being used as a fibre to make clothes it is often just called polyester. One advantage of PET is that it can be converted back into monomers and re-polymerised. In this way PET can be recycled.

QUESTIONS

1. SUPPLEMENT Nylon is an example of a condensation polymer. Explain how this type of polymer is different from an addition polymer.

2. SUPPLEMENT Explain why it is important that the monomers that join together to make nylon have reactive groups at each end of the molecule.

3. SUPPLEMENT Draw the repeat unit in the nylon polymer chain shown in Fig 11.31.

THE CHALLENGES OF RECYCLING PLASTICS

Plastics are made from polymers and are very difficult to dispose of. Most of them are not **biodegradable** – they cannot be decomposed by bacteria in the soil. Most waste plastic material is buried in landfill sites or burned, but burning plastics produces toxic fumes and landfill sites are filling up.

△ Fig. 11.32 Waste plastics are lightweight but very bulky.

Recycling is increasingly becoming normal practice. There is also an increase in the production of biodegradable plastics, which can be decomposed by bacteria in the soil. Biodegradable plastics are often called bioplastics and they are generally made from plants (such as bamboo or sugarcane) rather than fossil fuels.

However, there is growing evidence of waste plastic material accumulating in the oceans, which is causing a huge environmental challenge as it is very damaging to sea creatures.

△ Fig. 11.33 Waste plastic material is sorted during recycling.

Challenge Question: Biodegradable plastics can take between three and six months to decompose fully. Also, when some biodegradable plastics decompose in landfill sites they produce methane gas. Discuss the advantages and disadvantages of using biodegradable plastics.

End of topic checklist

Key terms

addition polymer, alkene, monomer, plastic, polymer

SUPPLEMENT condensation polymer, polyamide, repeat unit

During your study of this topic you should have learned:

○ That polymers are large molecules built up from many smaller molecules called monomers.

○ That poly(ethene) is formed from ethene monomers by addition polymerisation.

○ SUPPLEMENT How to identify repeat units in addition and condensation polymers.

○ SUPPLEMENT How to find the structure or repeat unit of an addition polymer.

○ SUPPLEMENT How to describe the differences between addition and condensation polymerisation.

○ SUPPLEMENT How to describe and draw the structures of nylon (a polyamide).

End of topic questions

1. State which of the following compounds can be used as a monomer to produce an addition polymer:

    A butene

    B butane

    C butanol

    D poly(ethene)

2. Define the following terms:

    a) *monomer*

    b) *polymer.*

3. **SUPPLEMENT** This question is about addition polymers.

    a) State what an *addition polymer* is.

    b) State what structural feature all monomers that form addition polymers have in common.

    c) Use displayed formulas to show how propene molecules react to form poly(propene).

    d) Explain why a polymer called poly(propane) does not exist.

4. **SUPPLEMENT** This question is about condensation polymers.

    a) Describe in what ways a condensation polymer is different to an addition polymer.

    b) i) Give the name of an example of a condensation polymer.

    ii) Describe a use of this condensation polymer.

Practice questions for Section 11

Example answer

Note: Practice questions, sample answers and comments have been written by the authors. The marks awarded for these questions indicate the level of detail required in the answers. In examinations, the way marks are awarded may be different. References to assessment and/or assessment preparation are the publisher's interpretation of the syllabus requirements and may not fully reflect the approach of Cambridge Assessment International Education.

Question 1

This question is about the following organic compounds:

A C_4H_{10} **B** C_2H_5OH **C** C_4H_8 **D** CH_3COOH

a) State which compound belongs to the alkene group.

 C ✓ ① (1)

b) State which compound is a saturated hydrocarbon.

 C ✗ (1)

c) State which compound will decolourise bromine water.

 C ✓ ① (1)

d) Give the names of the two products that are formed when compound A burns in a plentiful supply of air.

 carbon dioxide ✓ ① *and water* ✓ ① (2)

e) Give the names of two other substances that can be produced when compound A burns in a limited supply of air.

 carbon ✓ ① *and hydrogen* ✗ (2)

 Total ⑤/⑦

COMMENTS

a) Correct answer; alkenes are hydrocarbons with the general formula C_nH_{2n}.

b) The correct answer is A. Compound C is an alkene and is unsaturated; it contains a carbon-to-carbon double bond.

c) Correct answer. The decolourising of bromine water is a test for an alkene.

d) Correct answers. All hydrocarbons burn in a plentiful supply of oxygen forming carbon dioxide and water.

e) Carbon is correct and is responsible for the yellow flame characteristic of incomplete combustion. The missing substance is carbon monoxide, which is extremely poisonous.

Question 2 SUPPLEMENT

Petroleum is a mixture of different hydrocarbons. State which process is used to separate the petroleum into its different components:

A combustion

B fractional distillation

C cracking

D oxidation (1)

(Total 1 mark)

Question 3 SUPPLEMENT

The alkanes are a homologous series of saturated hydrocarbons.

a) State whether each of the following statements about the members of the alkane homologous series is TRUE or FALSE.

i) They have similar chemical properties. (1)

ii) They have the same displayed formula. (1)

iii) They have the same general formula. (1)

iv) They have the same physical properties. (1)

v) They have the same relative formula mass. (1)

b) Define the following terms:

i) hydrocarbon (1)

ii) saturated. (1)

c) The third member of the alkane homologous series is propane.

i) Give the molecular formula of propane. (1)

ii) Draw the displayed formula of propane. (2)

(Total 10 marks)

Question 4

Many useful substances are produced by the fractional distillation of crude oil.

a) Bitumen, diesel oil and gasoline are three fractions obtained from crude oil.

State which fraction has the following property:

i) the highest boiling point (1)

ii) molecules with the fewest carbon atoms. (1)

b) SUPPLEMENT Some long-chain hydrocarbons can be broken down into more useful products.

State the name of this process and describe how is it carried out. (3)

c) SUPPLEMENT Methane is a greenhouse gas. When methane is burned in a limited supply of air, carbon monoxide is formed.

i) Write a balanced equation for this reaction. (2)

ii) How does methane contribute to global warming? (2)

(Total 9 marks)

Question 5

Propene can be converted into a polymer called poly(propene).

propene

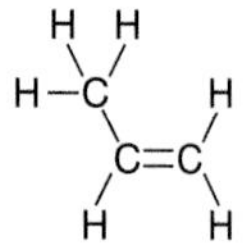

a) SUPPLEMENT State which homologous series propene belongs to. (1)

b) Give the general name given to an individual molecule that combines with other molecules to make a polymer. (1)

c) SUPPLEMENT Use the displayed formula of propene to show how three molecules link together to form part of the poly(propene) molecule. (2)

d) SUPPLEMENT Draw the repeat unit in poly(propene). (2)

e) State what type of polymer poly(propene) is. (1)

f) SUPPLEMENT Nylon is a different type of polymer to poly(propene).

i) State what type of polymer nylon is. (1)

ii) In terms of how it is made, describe how nylon is different from poly(propene). (1)

(Total 9 marks)

Question 6

Poly(ethene) is a plastic that is made by polymerising ethene, C_2H_4.

a) State which one of the following best describes the ethene molecules in this reaction. Draw a ring around the correct answer.

alcohols, alkanes, monomers, polymers, products (1)

b) The structure of ethane is shown below.

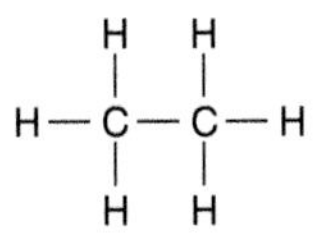

```
    H  H
    |  |
H — C — C — H
    |  |
    H  H
```

Explain, by referring to its bonding, why ethane cannot be polymerised. (1)

c) Draw the structure of ethene, showing all its atoms and bonds. (1)

d) SUPPLEMENT Ethene is obtained by cracking alkanes.

i) Explain the meaning of the term *cracking*. (1)

ii) State what conditions are needed to crack alkanes. (1)

iii) Copy and complete the equation for cracking decane, $C_{10}H_{22}$.

$C_{10}H_{22} \rightarrow C_2H_4 +$ _____ (1)

e) SUPPLEMENT Some oil companies crack the ethane produced when petroleum is distilled.

i) Copy and complete the equation for this reaction.

$C_2H_6 \rightarrow C_2H_4 +$ _____ (1)

ii) Describe the process of fractional distillation, which is used to separate the different fractions in petroleum. (2)

iii) State a use for the following petroleum fractions:

gasoline fraction (1)

refinery gas. (1)

(Total 11 marks)

One of the key aims of a chemistry course is to develop a range of experimental skills whilst making risk assessments in order to work safely. Many of the most important advances in chemistry have come about as a result of careful practical work. As this work underpins many of the other sections in this book, you may study parts of this section to link with the work in different sections. Alternatively, you may have covered some of this section in your earlier work in chemistry and so you could use this section for reference or revision.

In this section you will learn about experimental techniques, such as separation and purification techniques, ensuring you use the correct chemical terms. You will then learn about the analytical techniques which allow chemists to identify particular ions and gases.

Starting points

1. What apparatus would you use to measure the volume of a liquid? If you have named more than one piece of apparatus which one would give the most accurate reading?

2. You will have come across solutions in everyday life as well as in the laboratory. What is a solution?

3. Can you name an indicator that can be used to distinguish between an acid and an alkali? What results does this indicator give for an acid and an alkali?

4. Can you explain the terms 'anion' and 'cation'?

SYLLABUS SECTIONS COVERED

12.1 Experimental design

12.2 Acid–base titrations

12.3 Chromatography

12.4 Separation and purification

12.5 Identification of ions and gases

12
Experimental techniques and chemical analysis

△ The procedures and tests we do in a school laboratory are also used on a much larger scale in industry.

Experimental techniques

INTRODUCTION

Practical work is a very important part of studying chemistry. In your practical work you will need to develop your skills so that you can safely, correctly and methodically use and organise techniques, apparatus and materials. This involves being able to use appropriate apparatus for measurement to give readings to the required degree of accuracy. It is important to be able to use techniques that will determine the purity of a substance and, if necessary, techniques that can be used to purify mixtures of substances. You will always need to know when eye protection is needed. You will also need to avoid getting chemicals on your skin and avoid ingesting or inhaling anything that will cause you harm.

△ Fig. 12.1 Using the neutralisation method for a titration.

KNOWLEDGE CHECK

✓ Particular equipment is used for measuring time, temperature, mass and volume.
✓ Some substances are mixtures of a number of different components.

LEARNING OBJECTIVES

✓ Name appropriate apparatus for the measurement of time, temperature, mass and volume, including: stop-watches; thermometers; balances; burettes; volumetric pipettes; measuring cylinders; gas syringes.
✓ Describe a:
  a) solvent as a substance that dissolves a solute
  b) solute as a substance that is dissolved in a solvent
  c) solution as a mixture of one or more solutes dissolved in a solvent
  d) saturated solution as a solution containing the maximum concentration of a solute dissolved in the solvent at a specified temperature
  e) residue as a solid substance that remains after evaporation, distillation, filtration or any similar process
  f) filtrate as a liquid or solution that has passed through a filter.
✓ Describe an acid–base titration to include the use of a burette, volumetric pipette and suitable indicator.
✓ Describe how to identify the end-point of a titration using an indicator.
✓ Describe how paper chromatography is used to separate mixtures of soluble coloured substances, using a suitable solvent.
✓ Interpret simple chromatograms to identify: unknown substances by comparison with known substances; pure and impure substances.
✓ **SUPPLEMENT** State and use the equation for R_f:

$$R_f = \frac{\text{distance travelled by substance}}{\text{distance travelled by solvent}}$$

✓ Describe and explain methods of separation and purification using: a suitable solvent; filtration; crystallisation; simple distillation; fractional distillation.

✓ Suggest suitable separation and purification techniques, given information about the substances involved.

✓ Identify substances and assess their purity using melting point and boiling point information.

MEASUREMENT

In your study of chemistry you will carry out practical work. It is essential to select the right apparatus for the task, and to use it correctly so that your measurements are **accurate**. The chapter Developing experimental skills gives more details of selecting and using apparatus.

Time is measured with clocks, such as a wall clock. The clock should be accurate to about 1 second. A stop-watch is easier to use or you may be able to use your own wristwatch.

Temperature is measured using a thermometer. The range of the thermometer is commonly $-10\,°C$ to $+110\,°C$ with intervals of $1\,°C$.

Mass is measured with a balance or scales.

Volume of liquids can be measured with burettes, volumetric pipettes and measuring cylinders. Always make sure that you read the level of the liquid with your eye level with the meniscus.

Volume of gases can be measured with a gas syringe.

You will learn about the use of a burette, volumetric pipette and suitable indicator to identify the **end-point** of a **titration** between an acid and a base in Section 7, Acids, bases and salts.

△ Fig. 12.2 measuring equipment.

ACID-BASE TITRATION

Titration is an accurate experimental method of finding the volume of acid needed to neutralise a known volume of a solution of a base.

Using a burette rather than a measuring cylinder to measure the volume of the acid is more accurate because the smallest division of the scale of the burette is $0.1\ cm^3$. This means the volume measurement will be accurate to the nearest $0.1\ cm^3$. Pipettes are used to accurately measure a fixed volume of a solution, such as $25\ cm^3$. When the liquid is level with the filling line, the volume is precisely $25.0\ cm^3$.

The chapter Developing experimental skills gives examples of safe and accurate use of a burette and pipette in a titration.

Method:

1. 25.0 cm³ of the alkali is put into a conical flask using a pipette. The flask is placed on a white tile (to see colour changes more easily) and a few drops of an appropriate indicator (methyl orange) are added.

2. The burette is filled with an acid.

3. When the burette tap opens, the acid runs into the alkali in the flask and the flask is swirled.

4. Near the 'end-point' (neutralisation, when the indicator changes colour), the solution is added drop by drop until one drop changes the colour of the indicator from yellow to pink.

5. The volume of solution added from the burette is read and the whole process repeated until burette readings that are identical (or within 0.10 cm³) are obtained.

hydrochloric acid

burette

tap

sodium hydroxide solution of unknown concentration

white tile (to see colour change)

Δ Fig. 12.3 Titration apparatus.

PAPER CHROMATOGRAPHY

Paper **chromatography** is a way of separating **solutions** or **soluble** coloured substances that are mixed together.

For example, soluble black ink is often a **mixture** of different coloured inks. The diagrams in Fig. 12.4 show how paper chromatography is used to find the colours that make up a black ink.

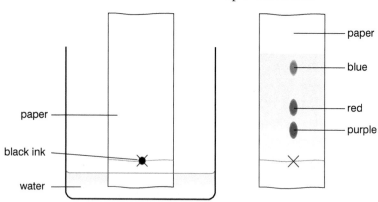

paper

black ink

water

paper

blue

red

purple

Δ Fig. 12.4 Paper chromatography separates a solution to find the colours in black ink. The left part of the diagram shows the paper before the inks have been separated, and the right part shows the paper after the inks have been separated.

A spot of ink is placed on the × mark and the paper is suspended in water. As the water rises up the paper, the different dyes travel different distances and so are separated on the **chromatogram**.

Paper chromatography can be used to identify what an unknown liquid is made of. This is called to interpreting a chromatogram.

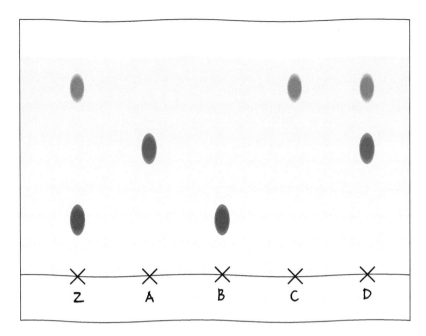

△ Fig. 12.5 A chromatogram.

The unknown liquid Z is compared with known liquids – in this case A to D.

Z must be made of B and C because the pattern of their dots matches the pattern shown by Z. If Z only produced one colour then it would be a **pure** substance. The chromatogram shows that Z is an **impure** substance as it is made of a mixture of two substances.

△ Fig. 12.6 A piece of filter paper is marked with black ink and dipped into water in a beaker.

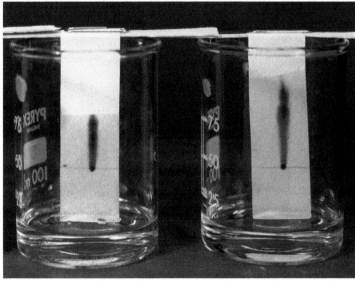

△ Fig. 12.7 After a few minutes the chromatogram has been created by the action of the water on the ink.

R_f value

Substances can also be identified using chromatography by measuring what is known as the R_f **value** on the filter paper. The R_f **value** for a particular substance compares the distance the substance has travelled up the filter paper with the distance travelled by the **solvent**. The R_f value can be calculated using the following formula:

$$R_f = \frac{\text{distance travelled by substance}}{\text{distance travelled by solvent}}$$

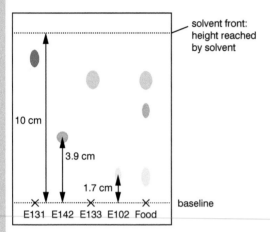

Δ Fig. 12.8 The R_f value for the food additive E102 is 0.17.

As the solvent will always travel further than the substance, R_f values will always be less than 1.

CRITERIA FOR PURITY

The purity of solids and liquids

It is very important that manufactured foods and medicines contain only the substances the manufacturers want in them – that is, they must not contain any contaminants or impurities.

The simplest way of checking the purity of solids and liquids is using heat to find the temperature at which they melt (melting point) or boil (boiling point).

An impure solid will have a lower melting point than the pure solid.

A liquid containing a dissolved solid (solute) will have a higher boiling point than the pure solvent.

The best examples to use to remember these facts are water and ice:

- Pure water boils at 100 °C; salted water for cooking vegetables boils at about 102 °C.
- Pure ice melts at 0 °C; ice with salt added to it melts at about –4 °C.

1. The start line, or baseline, in chromatography should be drawn in pencil. Explain why.

2. In a chromatography experiment, explain why the solvent level in the beaker must be below the baseline.

3. In a chromatography experiment to compare the dyes in two different inks, one of the inks does not move at all from the baseline. Suggest a reason for this.

4. A sample of water contains some dissolved impurities. Suggest what you would expect the boiling point of the sample to be.

5. SUPPLEMENT Look at the diagram in Fig. 12.8. Explain why the R_f value for the food additive E102 is 0.17.

METHODS OF PURIFICATION

Techniques for separating mixtures of solids and liquids or mixtures of liquids rely on finding different properties of the substances that make up the mixture.

Separating soluble and insoluble solids

The method is:

1. Add a **solvent** that the required solid is **soluble** in, and dissolve it to form a **solution**. The solid is a **solute**.

2. Filter the mixture to remove the insoluble impurity, known as the **residue**. The liquid or solution which passes through the filter paper is called the **filtrate**.

3. Heat the filtrate to remove some solvent and leave it to **crystallise**.

4. Filter off the crystals, wash with a small amount of cold solvent and dry them – this is the pure solid.

Note: A solution which has dissolved the maximum amount of a solute at a given temperature is known as a **saturated solution**.

An example of using this technique would be separating salt from 'rock salt' (the impure form of sodium chloride). Water is added to dissolve the salt but leave the other solids undissolved. Filter off the insoluble impurities, warm the salt solution and leave it to crystallise to form salt crystals.

△ Fig. 12.9 Filtration of copper(II) hydroxide.

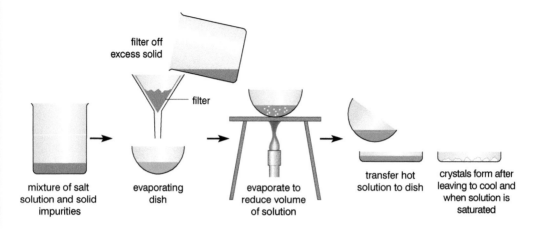

Δ Fig. 12.10 Separating impurities in rock salt.

Separating and purifying liquids

A pure liquid contains only one substance, for example pure water contains only molecules of water and is not a mixture with

- another liquid,
- any dissolved substances,
- any undissolved solids

There are three methods of purifying an impure liquid:

1. Liquids contaminated with soluble solids dissolved in them.

The method is simple **distillation**.

The solution is heated, the solvent boils and turns into a vapour. It is condensed back to the pure liquid and collected.

This is the technique used in desalination plants, which produce pure drinking water from sea water. The solids are left behind after boiling off the water.

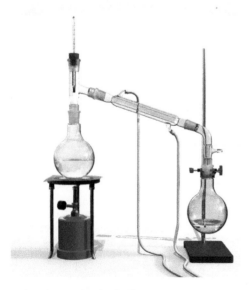

Δ Fig.12.11 A simple distillation apparatus.

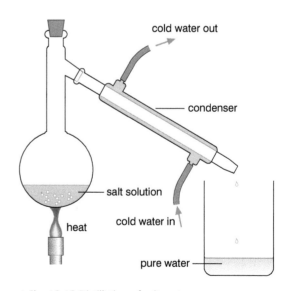

Δ Fig. 12.12 Distillation of salt water.

2. Liquids contaminated with other liquids.

In this case the technique is **fractional distillation**, which uses the difference in boiling points of the different liquids mixed together.

The mixture is boiled, and the liquid with the lowest boiling point turns to a vapour first, rises up the fractionating column and is condensed back to liquid in the condenser. The next lowest boiling point liquid comes off, and so on until all the liquids have been separated. You can identify the fraction you want to collect by the temperature reading on the thermometer. The fractionating column increases the purity of the distilled product by reducing the amount of other substances in the vapour when it condenses.

Fractional distillation is the method used to separate the mixture of hydrocarbons in crude oil. This method can be found in Section 11 Organic chemistry. It is not necessary to learn this as it is not on the syllabus, but fractional distillation is also used when collecting ethanol (alcohol) from a fermentation mixture.

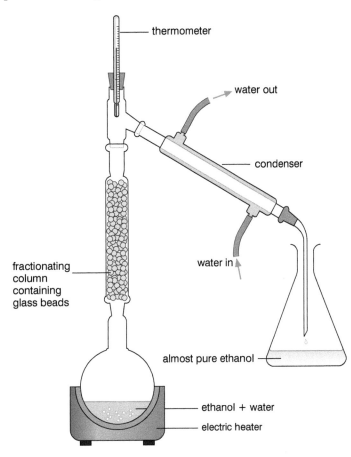

Δ Fig. 12.13 Apparatus for fractional distillation of an alcohol/water mixture.

3. Liquids contaminated with **insoluble** solids can be separated by **filtration** as shown in Fig 12.9.

QUESTIONS

1. Explain the term *'solvent'*.

2. Explain the term *'soluble'*.

3. Suggest what method you would use to separate a pure liquid from a solution of a solid and the liquid.

4. To separate two liquids by fractional distillation they must have different:

   A melting points

   B boiling points

   C colours

   D viscosities.

Key terms

accurate, chromatogram, crystallisation, desalination, end-point, filtrate, filtration, fractional distillation, impure, insoluble, meniscus, mixture, paper chromatography, pure, residue, saturated solution, soluble, solute, solution, solvent, titration

SUPPLEMENT R_f value

During your study of this topic you should have learned:

○ Name appropriate apparatus for the measurement of time, temperature, mass and volume, including:

- Stopwatches, thermometers, balances, burettes, volumetric pipettes, measuring cylinders, gas syringes.

○ How to describe a solvent, solute, solution, saturated solution, residue and filtrate.

○ How to describe an acid–base titration, including identification of the end-point.

○ About the technique of paper chromatography.

○ How to interpret simple chromatograms.

○ About a range of suitable separation techniques involving dissolving in a solvent, filtration, crystallisation, simple distillation and fractional distillation.

○ How to suggest suitable purification techniques given information about the substances involved.

○ How to identify substances and assess their purity from melting point and boiling point information.

○ **SUPPLEMENT** How to interpret simple chromatograms, including the use of R_f values.

End of topic questions

1. State which one of the following mixtures could be separated by filtration:

    A sand from a sand/water mixture

    B petrol from a petrol/diesel mixture

    C pure water from salt solution

    D water from a water-based ink

2. You are provided with four samples of black water-soluble inks.
    Two of the ink samples are identical. Describe how you would use
    paper chromatography to identify which two ink samples are the same.

3. You are trying to separate the dyes in a sample of ink using paper
    chromatography. You set up the apparatus as shown in Fig. 12.4.
    After 20 minutes the black spot is unchanged and the water has risen
    nearly to the top of the filter paper.

    a) Suggest a reason why the black spot has remained unchanged.

    b) Suggest what you could change that might lead to a successful
    separation of the dyes.

4. Pure ethanol has a boiling point at normal pressure of 78 °C. Predict
    what temperature a sample of ethanol contaminated with sugar might
    boil at.

5. Predict what effect impurities will have on the melting point of ice.

6. In the fractional distillation of ethanol and water (see Fig. 12.13),
    explain why the ethanol vapour condenses in the condenser.

7. Describe how you would produce crystals of sodium chloride
    from a sodium chloride solution.

8. SUPPLEMENT Look at the chromatogram produced when testing four food colouring compounds A, B, C and D.

a) State which compound has the largest R_f value.

b) State which compound has the smallest R_f value.

c) Estimate the R_f value for compound C. Explain how you made the estimate.

d) Explain why all R_f values are less than 1.0.

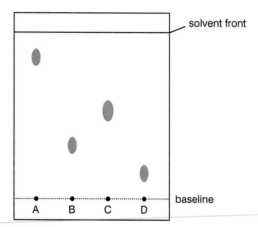

Identification of ions and gases

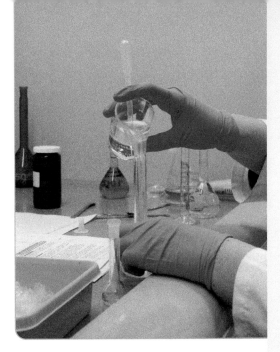

△ Fig. 12.14 A chemist performs a chemical test on a substance in her pharmaceutical laboratory.

INTRODUCTION

It is important to be able to analyse different substances and identify the different elements or components. The techniques used today are fairly sophisticated but many of them are based on simple laboratory tests. With improved understanding of the beneficial and harmful properties of chemical substances, it has become more and more important to identify metals and non-metals in chemical processes and in the environment.

KNOWLEDGE CHECK

✓ Ionic compounds contain a metal combined with a non-metal.
✓ An anion is a negatively charged ion.
✓ A cation is a positively charged ion.

LEARNING OBJECTIVES

✓ Describe tests to identify the anions:
  a) carbonate, CO_3^{2-}, by reaction with dilute acid and then testing for carbon dioxide gas
  b) chloride, Cl^-, bromide, Br^-, and iodide, I^-, by acidifying with dilute nitric acid then adding aqueous silver nitrate
  c) nitrate, NO_3^-, reduction with aluminium foil and aqueous sodium hydroxide and then testing for ammonia gas
  d) sulfate, SO_4^{2-}, by acidifying with dilute nitric acid and then adding aqueous barium nitrate.
✓ Describe tests using aqueous sodium hydroxide and aqueous ammonia to identify the aqueous cations:
  a) ammonium (NH_4^+)
  b) calcium (Ca^{2+})
  c) copper(II) (Cu^{2+})
  d) iron(II) (Fe^{2+})
  e) iron(III) (Fe^{3+})
  f) zinc (Zn^{2+}).
    (formulas of complex ions are not required).
✓ Describe tests to identify the gases:
  a) ammonia, NH_3, using damp red litmus paper
  b) carbon dioxide, CO_2, using limewater
  c) chlorine, Cl_2, using damp litmus paper
  d) hydrogen, H_2, using a lighted splint
  e) oxygen, O_2, using a glowing splint.
✓ Describe the use of a flame test to identify the cations: lithium (Li^+), sodium (Na^+), potassium (K^+), copper(II) (Cu^{2+}).

Ions of metals are **cations** – positive ions – and are found in ionic compounds. There are two ways of identifying metal cations:

- *either* from solids of the compound
- *or* from solutions of the compound.

Flame tests

In a flame test, a piece of nichrome wire is dipped into concentrated hydrochloric acid, then into the solid compound, and then into a blue Bunsen flame. The colour seen in the flame identifies the metal ion in the compound.

Name of ion	Formula of ion	Colour seen in flame
lithium	Li^+	red
sodium	Na^+	yellow
potassium	K^+	lilac
copper	Cu^{2+}	blue-green

Δ Table 12.1 Colours of ions in a flame.

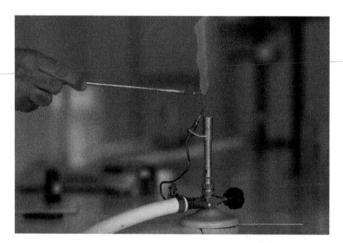

Δ Fig. 12.15 The colour of the flame can be used to identify the metal ions present.

Tests on solutions in water

Metal ions are found in ionic compounds, so most of them will dissolve in water to form solutions. These solutions can be tested with aqueous sodium hydroxide solution and aqueous ammonia to identify the aqueous cation by the colour of a precipitate formed in the reaction.

Name of ion in solution	Formula	Result
Calcium	$Ca^{2+}(aq)$	White precipitate, insoluble in excess
Copper(II)	$Cu^{2+}(aq)$	Light blue precipitate formed, insoluble in excess
Iron(II)	$Fe^{2+}(aq)$	Green precipitate formed, insoluble in excess, precipitate turns brown near surface on standing
Iron(III)	$Fe^{3+}(aq)$	Red-brown precipitate, insoluble in excess
Zinc	$Zn^{2+}(aq)$	White precipitate, soluble in excess, giving a colourless solution

△ Table 12.2 Tests for identifying cations by adding sodium hydroxide solution to excess.

Note:

Similar results can be obtained if these reactions are performed using ammonia solution instead of sodium hydroxide solution. However, there is a noticeable difference in the case of copper(II) ions. At first, as with sodium hydroxide solution, a light blue precipitate is formed, but then as excess ammonia solution is added the precipitate dissolves to form a dark blue solution.

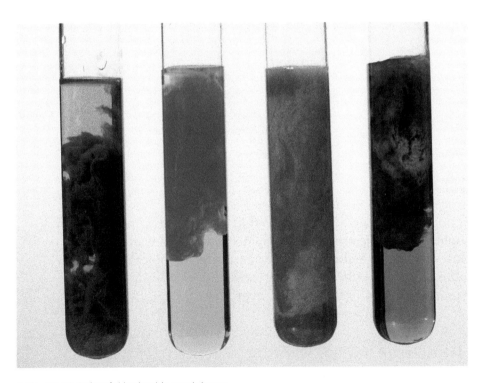

△ Fig. 12.16 Colourful hydroxide precipitates.

The reactions in the table can be represented by **ionic equations**. For example:

$$Ca^{2+}(aq) + 2OH^-(aq) \rightarrow Ca(OH)_2(s)$$
$\qquad$ white precipitate
$$Cu^{2+}(aq) + 2OH^-(aq) \rightarrow Cu(OH)_2(s)$$
$\qquad$ light blue precipitate
$$Fe^{3+}(aq) + 3OH^-(aq) \rightarrow Fe(OH)_3(s)$$
$\qquad$ red-brown precipitate

QUESTIONS

1. **a)** Describe what you would observe if you added sodium hydroxide solution to a solution of calcium chloride.

   b) Suggest in what way the observations would be different if zinc chloride solution were used instead of calcium chloride solution.

2. Describe what test can be used to distinguish between Fe^{2+} and Fe^{3+} ions. Give the result of the test with each ion.

IDENTIFYING AMMONIUM IONS, NH_4^+

The test for the ammonium ion is shown in Fig. 12.17.

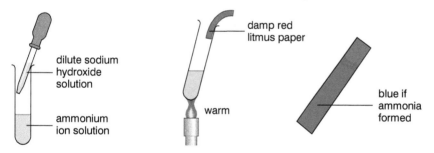

△ Fig. 12.17 Test for the ammonium ion NH_4^+.

The suspected ammonium compound is dissolved in water in a test-tube and a few drops of dilute sodium hydroxide are added. The mixture is then warmed over a Bunsen burner and some damp red litmus paper (or universal indicator paper) is placed in the mouth of the test-tube. A colour change in the indicator to blue (alkaline) shows that an ammonium compound is present.

IDENTIFYING ANIONS

Negative ions (**anions**) can be tested as solids or in solution.

Testing for anions in solids or solutions

The following test for anions in solids applies only to **carbonates**.

Dilute hydrochloric acid is added to the solid, and any gas produced is passed through limewater. If the limewater goes cloudy/milky, the solid contains a carbonate.

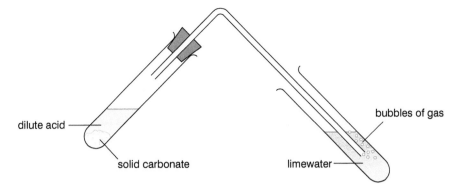

dilute acid

solid carbonate

bubbles of gas

limewater

Δ Fig. 12.18 Testing for anions in solids.

This reaction is as follows:

acid + carbonate → a salt + water + carbon dioxide

$2HCl(aq) + Na_2CO_3(s) \rightarrow 2NaCl(aq) + H_2O(l) + CO_2(g)$

$2HCl(aq) + ZnCO_3(s) \rightarrow ZnCl_2(aq) + H_2O(l) + CO_2(g)$

SUPPLEMENT

The reaction between an acid and a carbonate can be represented by the following ionic equation:

$CO_3^{2-}(s) + 2H^+(aq) \rightarrow CO_2(g) + H_2O(l)$

Many ionic compounds are soluble in water, and so they form solutions that contain anions.

The tests and results used to identify some other common anions are shown in the table.

Name of ion	Formula	Test	Result
Carbonate	CO_3^{2-}	Add dilute acid, then test for carbon dioxide gas	Effervescence, carbon dioxide produced
Chloride	$Cl^-(aq)$	Add: 1. Dilute nitric acid 2. Silver nitrate solution	White precipitate
Sulfate	$SO_4^{2-}(aq)$	Add: 1. Dilute hydrochloric acid or nitric acid 2. Barium chloride or nitrate solution	White precipitate
Nitrate	$NO_3^-(aq)$	1. Add sodium hydroxide solution and warm 2. Add aluminium foil or powder 3. Test any gas produced with damp red litmus paper	Ammonia produced

Δ Table 12.3 Tests for anions.

△ Fig. 12.19 Test for chloride: white precipitate.

SUPPLEMENT

This reaction can be represented by an ionic equation:

$Ag^+(aq) + Cl^-(aq) \rightarrow AgCl(s)$
      white precipitate

QUESTIONS

1. Describe how you would test for an ammonium compound. Give the result of the test.

2. When a carbonate is reacted with dilute hydrochloric acid, a gas is given off.

  a) State the name of the gas.

  b) Describe the test for the gas. Give the result of the test.

3. Sodium hydroxide solution is added to solution X and a reddish-brown precipitate is formed. State what metal ion was present in solution X.

4. A mixture of dilute nitric acid and silver nitrate solution is added to solution Y in a test-tube. A white precipitate forms. State what anion is present in solution Y.

5. A forensic scientist has been provided with a small sample of a blue compound which is suspected to be copper(II) sulfate, and a white compound that is suspected to be sodium carbonate. Devise a series of tests that could be followed to identify the ions. Indicate in your plan the expected results if the samples are to be positively identified.

6. Metallic ions in solution can be identified using sodium hydroxide solution. Sodium hydroxide is useful because it forms coloured precipitates with many metallic ions, although it will form white precipitates with others.

   a) Copy and complete the table with the names of two cations that form white precipitates and three cations that give coloured precipitates.

Name of cation	Colour of precipitate

   b) SUPPLEMENT Halide ions can be identified using nitric acid and silver nitrate. Give a general ionic equation to show the formation of the precipitate of the halide ion.

IDENTIFYING GASES

Many chemical reactions produce a gas as one of the products. Identifying the gas is often a step towards identifying the compound that produced it in the reaction.

Gas	Formula	Test	Result of test
Hydrogen	H_2	Put in a lighted splint (a flame)	'Pops' with a lighted splint
Oxygen	O_2	Put in a glowing splint	Relights a glowing splint
Carbon dioxide	CO_2	Pass gas through limewater	Turns limewater milky
Chlorine	Cl_2	Put in a piece of damp blue litmus paper or universal indicator paper	Bleaches damp litmus paper
Ammonia	NH_3	Put in a piece of damp red litmus paper or universal indicator paper	Turns damp red litmus paper blue

△ Table 12.4 Tests for gases.

It is important in practical work where you are identifying gases to remember that small amounts of chlorine and ammonia can cause respiratory distress. Therefore it is very important that the laboratory is well ventilated and only small amounts of gas are produced.

Carbon dioxide: cloudiness with limewater is caused by insoluble calcium carbonate. If carbon dioxide continues to be passed through, the cloudiness disappears: $CaCO_3(s)$ is changed to soluble calcium hydrogencarbonate, $Ca(HCO_3)_2(aq)$.

Chlorine: the gas is acidic, but also a bleaching agent.

Ammonia: the only common basic/alkaline gas.

QUESTIONS

1. Describe how to test for ammonia gas.

2. Give the name of a gas that supports combustion.

3. Give the name of a gas that acts as a bleach.

End of topic checklist

Key terms

anion, carbonate, cation, precipitate

SUPPLEMENT ionic equation

During your study of this topic you should have learned:

○ How to carry out tests for these anions:

- carbonate by reaction with dilute acid and then testing for carbon dioxide gas
- chloride, bromide and iodide by reaction under acidic conditions with aqueous silver nitrate
- nitrate by adding aqueous sodium hydroxide, then reduction with aluminium and testing for ammonia gas
- sulfate by reaction under acidic conditions with aqueous barium ions

○ How to carry out tests for these aqueous cations: ammonium, calcium, copper(II), iron(II), iron(III) and zinc using aqueous sodium hydroxide and aqueous ammonia.

○ How to carry out tests for these gases:

- ammonia using damp red litmus paper
- carbon dioxide using limewater
- chlorine using damp litmus paper
- hydrogen using a lighted splint
- oxygen using a glowing splint

○ How to use a flame test to identify these cations: lithium, sodium, potassium and copper(II).

1. State which of the following solutions contains Fe^{3+} ions:

   A The solution forms a pale blue precipitate when sodium hydroxide solution is added.

   B The solution forms no precipitate when sodium hydroxide solution is added, but produces a strong-smelling, alkaline gas when the mixture is heated.

   C The solution forms an orange-brown precipitate when sodium hydroxide solution is added.

   D The solution forms a green precipitate when sodium hydroxide solution is added.

2. Copy and complete the table about the identification of gases.

Gas	Test	Observations
Chlorine	Damp universal indicator paper	
	Bubble through limewater	White precipitate or suspension forms
Hydrogen		Burns with a 'pop'

3. A white powder is labelled 'lithium carbonate'. Describe a test you could do to prove it was a carbonate.

4. Describe how you would test a solid to identify the presence of each of the ions shown below.

   a) the sulfate ion, SO_4^{2-}

   b) the iodide ion, I^-

   c) the nitrate ion, NO_3^-

5. **SUPPLEMENT** Write ionic equations for the following reactions:

   a) between copper(II) sulfate and sodium hydroxide solution

   b) between sodium carbonate and dilute hydrochloric acid.

COMMENTS

a) Correct answer. Iron and chromium are other possible answers.

b) A very good answer. Cleaning the nichrome wire is a point often missed.

c) Incorrect answer. Potassium gives a lilac (light purple) flame colour. Lithium gives a red flame colour.

d) i) Correct answer.

ii) Correct answer.

e) Incorrect answer. The formula for potassium carbonate is K_2CO_3 and for lithium carbonate is Li_2CO_3.

Practice questions for Section 12

Example answer

Note: Practice questions, sample answers and comments have been written by the authors. The marks awarded for these questions indicate the level of detail required in the answers. In examinations, the way marks are awarded may be different. References to assessment and/or assessment preparation are the publisher's interpretation of the syllabus requirements and may not fully reflect the approach of Cambridge Assessment International Education.

Question 1

A student is provided with a sample of an ionic solid A.

a) The solid is white. Name a metal ion that is not present in solid A and give a reason for your choice.

The copper ion (Cu^{2+}). Copper is a transition metal and forms coloured compounds, usually blue. ✔ ② (2)

b) The student undertakes a flame test on solid A. Describe how to perform the flame test.

Take a nichrome wire and heat in a blue Bunsen flame to remove any solid previously tested. Dip the wire into concentrated hydrochloric acid, then into solid A and then into a blue Bunsen flame. ✔ ③ (3)

c) The flame colour is red. State what metal ion is present in solid A.

The potassium ion is present. ✗ (1)

d) Dilute hydrochloric acid is added to solid A and a gas is produced which turned limewater cloudy.

i) State the name of the gas produced in this reaction.

Carbon dioxide ✔ ① (1)

ii) State which anion is present in solid A.

the carbonate ion ✔ ① (1)

e) Give the name and formula of solid A.

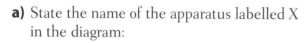

Potassium carbonate, KCO₃ ✗ .... (1)

(Total 9 marks)

⑦⁄₉

Question 2

A student tested a sample of sea water for dissolved solids. The apparatus used is shown on the right.

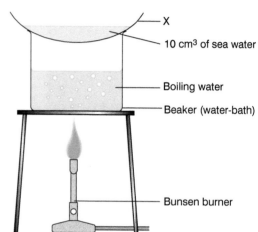

a) State the name of the apparatus labelled X in the diagram:

   A Boiling tube

   B Condenser

   C Watch glass

   D Funnel   (1)

b) The student did the test 4 times and then each time calculated the mass of the solid formed on apparatus X after heating. The results are shown in the table:

	Test 1	Test 2	Test 3	Test 4
Mass of solid (g)	0.12	0.29	0.16	0.14

  i) Identify which one of the results is anomalous.   (1)

  ii) Calculate the mean mass of solid formed. Do not include the anomalous result in your calculation. Give your answer to 2 significant figures.   (2)

c) The student then distilled a sample of the water using the apparatus shown on the right:

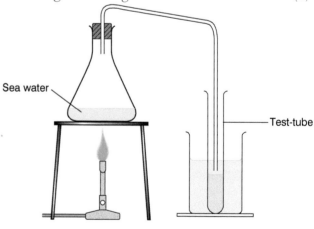

  i) Identify what change of state is occurring in the flask.   (1)

  ii) Identify what change of state is occurring in the test-tube.   (1)

  iii) Describe how the water in the test-tube will be different from the water in the conical flask.   (1)

  iv) Explain why distillation is not commonly used in the purification of domestic drinking water.   (1)

d) River water is filtered and then sterilised to make drinking water. Explain what is used in these two processes and how each process changes the river water.

i) Filtering (2)

ii) Sterilising (2)

(Total 12 marks)

Question 3 SUPPLEMENT

A student uses chromatography to identify the number of soluble components in a sample of black ink.

a) Describe how to perform the chromatography experiment. Include a diagram as part of your answer. (3)

b) Describe how the student obtained the R_f values for each component.

Component	R_f value
A	0.75
B	0.25

Draw and label the chromatogram you would expect in this experiment. (2)

c) Describe the appearance of the chromatogram the student would have obtained. (1)

(Total 6 marks)

Question 4

Complete the table below, which shows the common tests used for identifying some gases.

Gas	Formula	Test	Result of test
		Put in a glowing splint	Splint relights, producing a flame
Chlorine	Cl_2		
Ammonia		Put in a piece of damp red litmus or universal indicator paper	
		Put in a piece of filter paper soaked in potassium manganate(VII) solution	Filter paper changes from purple to colourless

(8)

(Total 8 marks)

Question 5

Complete the table below, which shows some of the identification tests for common anions.

Name of ion	Formula	Test	Observation
Chloride	$Cl^-(aq)$		
Sulfate			White precipitate
		1. Add sodium hydroxide solution and warm 2. Add aluminium foil or powder 3. Test any gas produced with damp red litmus paper	Red litmus paper goes blue (ammonia gas is produced)

(6)

(Total 6 marks)

Developing experimental skills

The information in this section is based on the Cambridge International syllabus. You should always refer to the appropriate syllabus document for the year of examination to confirm the details and for more information. The syllabus document is available on the Cambridge International website at www.cambridgeinternational.org.

INTRODUCTION

As part of your chemistry study of the Cambridge IGCSE Co-ordinated Sciences course, you will develop practical skills and have to carry out investigative work in science. This will either be done as coursework, a practical test or a written paper, depending on your school.

This section provides guidance on carrying out an investigation.

The experimental and investigative skills are divided as follows:

1. Apparatus and techniques
2. Plan an experiment
3. Make observations and measurements
4. Interpret observations and data
5. Evaluate methods

1. APPARATUS AND TECHNIQUES

Learning objective: to demonstrate and describe appropriate experimental and investigative methods, including safe and skilful practical techniques.

Questions to ask:

How shall I use the equipment safely to minimise the risks – what are my safety precautions?

✓ When writing a Risk Assessment, investigators need to be careful to check that they've matched the hazard with the technique used and with the concentration of a chemical used. Many acids, for instance, are corrosive in higher concentrations, but are likely to be irritants or of low hazard in the concentration used when working in biology or chemistry experiments.

✓ Don't forget to consider the hazards associated with all the chemicals and biological materials, even if these are very low.

✓ You may be asked to describe the precautions taken when carrying out an investigation.

How much detail should I give in my description?

✓ You need to give enough detail so that someone else who has not done the experiment would be able to carry it out to reproduce your results.

How should I use the equipment to give me the precision I need?

✓ You should know how to read the scales on the measuring equipment you are using.

✓ You need to show that you are aware of the precision needed.

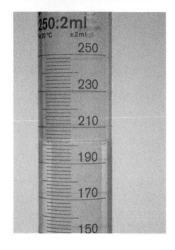

◁ Fig. 13.1 The volume of liquid in a measuring cylinder must be read to the bottom of the meniscus.

EXAMPLE 1

This is an extract from a student's notebook. It describes how she carried out a titration experiment using sulfuric acid and potassium hydroxide solutions, with methyl orange as an indicator.

$$2KOH(aq) + H_2SO_4(aq) \rightarrow K_2SO_4(aq) + 2H_2O(l)$$

What are my safety precautions?

Equipment

I will be using a pipette, burette and conical flask all made from glass so I will need to handle them carefully and, in particular, clamp the burette carefully and make sure the pipette does not roll off the bench when I am not using it. I will also be careful when attaching the pipette to the pipette filler so that the pipette does not break.

Chemicals

I have looked up the hazards

Sulfuric acid 0.1M: LOW HAZARD

Potassium hydroxide (0.1M approx.): IRRITANT

COMMENT

The student has used a data source to look up the chemical hazards. There is no risk assessment given for methyl orange. Methyl orange indicator solution is flammable, and toxic if swallowed. It should not be disposed of down a drain.

I will need to handle the chemicals carefully, have a damp cloth ready to wipe up any spills and wear eye protection.

COMMENT

The student has suggested some sensible precautions.

The student's method is given below.

A pipette and a burette were carefully washed, making sure they were drained after washing.

25.00 cm³ of the potassium hydroxide solution was measured in a pipette and transferred to a clean conical flask. 3 drops of methyl orange indicator were then added.

The burette was filled with the sulfuric acid solution, making sure that there were no air bubbles in the jet of the burette. The first reading of the volume of acid in the burette was taken.

The acid was added to the alkali in the conical flask, swirling the flask all the time. When the indicator colour was close to changing, the acid was added drop by drop until the colour changed. The second reading on the burette was taken.

The whole procedure was repeated twice, making sure that between each experiment the conical flask was washed carefully. The results are shown in the table.

COMMENT

The method is well written and detailed. The first point could have been improved if she had said that the burette and pipette had been washed with distilled water first and then the chemical to be used (burette – acid; pipette – alkali).

Precision and accuracy. Some examples from the notebook are:

'25.00 cm³ of the potassium hydroxide solution'

The student has appreciated the accuracy a titration can achieve.

'making sure that there were no air bubbles in the jet of the burette'

An air bubble could easily lead to an inaccurate measurement.

'the acid was added drop by drop until the colour changed'

Again the student tried to get accuracy to within one drop ($\pm\ 0.05\ \text{cm}^3$).

2. PLAN AN EXPERIMENT

Learning objective: to devise and plan investigations, drawing on scientific knowledge and understanding in selecting appropriate techniques.

Questions to ask:

What do I already know about the area of science I am investigating and how can I use this knowledge and understanding to help me with my plan?

✓ Think about what you have already learned and any investigations you have already done that are relevant to this investigation.

✓ List the factors that might affect the process you are investigating.

What is the best method or technique to use?

✓ Think about whether you can use or adapt a method that you have already used.

✓ A method, and the measuring instruments, must be able to produce **valid** measurements. A measurement is valid if it measures what it is supposed to be measuring.

You will make a decision as to which technique to use based on:

✓ The accuracy and precision of the results required; investigators might require results that are as accurate and precise as possible but if you are making a quick comparison, or a preliminary test to check a range over which results should be collected, a high level of accuracy and precision may not be required.

✓ The simplicity or difficulty of the techniques available, or the equipment required; is this expensive, for instance?

✓ The scale, for example, using standard laboratory equipment or on a micro-scale, which may give results in a shorter time period.

✓ The time available to do the investigation.

✓ Health and safety considerations.

What am I going to measure?

✓ The factor you are investigating is called the **independent variable**. A **dependent variable** is affected or changed by the independent variable that you select.

✓ You need to choose a range of measurements that will be enough to allow you to plot a graph of your results and so find out the pattern in your results.

✓ You might be asked to explain why you have chosen your range rather than a lower or higher range.

How am I going to control the other variables?

✓ These are **control variables**. Some of these may be difficult to control.

✓ You must decide how you are going to control any other variables in the investigation and so ensure that you are using a fair test and that any conclusions you draw are valid.

✓ You may also need to decide on the concentration or combination of reactants.

What equipment is suitable and will give me the accuracy and precision I need?

✓ The **accuracy** of a measurement is how close it is to its true value.

✓ **Precision** is related to the smallest scale division on the measuring instrument that you are using; for example, when measuring a distance, a ruler marked in millimetres will give greater precision than one divided into centimetres only.

✓ A set of precise measurements also refers to measurements that have very little spread about the mean value.

✓ You need to be sensible about selecting your devices and make a judgement about the degree of precision. Think about what is the least precise variable you are measuring and choose suitable measuring devices. There is no point having instruments that are much more precise than the precision you can measure the variable to.

What are the potential hazards of the equipment and technique I will be using and how can I reduce the risks associated with these hazards?

✓ Be prepared to suggest safety precautions when presented with details of an investigation.

✓ You can find out about hazards associated with reactants using CLEAPSS Student Safety Sheets or a similar resource.
Visit www.cleapss.org.uk.

EXAMPLE 2

You have to compare the effect of surface area on the rate of reaction between magnesium and dilute hydrochloric acid. You will be provided with 10 cm of magnesium ribbon and 0.1 mol / dm³ hydrochloric acid.

The student has written:

I will use the following method:

1. *I will look up the risk assessments for magnesium and dilute (0.1 mol / dm³) hydrochloric acid using the reference materials we have in the laboratory.*

2. *I will cut the magnesium ribbon into two 5 cm lengths and then use scissors to cut one of the 5 cm strips into very small pieces.*

3. *I will use a 100 cm³ measuring cylinder to measure 25 cm³ of the 0.1 mol / dm³ hydrochloric acid into a beaker.*

4. *I will add the 5 cm strip of magnesium ribbon to the acid and time how long it takes to dissolve completely.*

5. *I will then repeat the experiment using the very small pieces of magnesium ribbon.*

6. *I will use the times taken to dissolve to contrast the rates of reaction.*

The student has correctly demonstrated the importance of undertaking a risk assessment and has chosen an appropriate technique to compare the two reactions. The measuring cylinder and beaker are appropriate, but the student has not mentioned how time is to be measured – a stop watch would have been appropriate. The approach could be improved by saying how the resultant solution should be disposed of at the end of each reaction and how the beaker is washed (distilled water) and dried before the second experiment is performed.

3. MAKE OBSERVATIONS AND MEASUREMENTS

Learning objective: to make observations and measurements with appropriate precision, record these methodically, and present them in a suitable form.

Questions to ask:

How many different measurements or observations do I need to take?

✓ It is usual to repeat an experiment to take more than one measurement. If an investigator takes just one measurement, this may not be typical of what would normally happen when the experiment was carried out.

✓ When repeat readings are consistent, they are said to be **repeatable**.

Do I need to repeat any measurements or observations that are anomalous?

✓ An **anomalous result** or **outlier** is a result that is not consistent with other results.

✓ You want to be sure that a single result is accurate so you will need to repeat the experiment until you get close agreement in the results you obtain.

✓ If an investigator has made repeat measurements, they would normally use these to calculate the arithmetical mean (or just mean or average) of these data to give a more accurate result. You calculate the mean by adding together all the measurements, and dividing by the number of measurements. Be careful, though: anomalous results should not be included when taking averages.

✓ Anomalous results might be the consequence of an error made in measurement. But sometimes outliers are genuine results. If you think an outlier has been introduced by careless practical work, you should omit it when calculating the mean. But you should examine possible reasons carefully before just leaving it out.

✓ You are taking a number of readings in order to see a changing pattern. For example, measuring the volume of gas produced in a reaction every 10 seconds for 2 minutes (so 12 different readings).

✓ You can often pick an anomalous reading out from a results table (or a graph if all the data points have been plotted, as well as the mean, to show the range of data). It may be a good idea to repeat this part of the practical again, but it's not necessary if the results show good consistency.

✓ If you are confident that you can draw a line of best fit through most of the points, it is not necessary to repeat any measurements that are obviously inaccurate. If, however, the pattern is not clear enough to draw a graph then readings will need to be repeated.

How should I record my measurements or observations – is a table the best way? What headings and units should I use?

✓ A table is often the best way to record results.

✓ Headings should be clear.

✓ If a table contains numerical data, do not forget to include units; data are meaningless without them.

✓ The units should be the same as those that are on the measuring equipment you are using.

✓ Sometimes you are recording observations that are not quantities. Putting observations in a table with headings is a good way of presenting this information.

EXAMPLE 3

The student from Example 1 has recorded the results in a table, as shown in Table 13.1. In this case she needs to get two results within 0.1 cm³ of each other and so has had to do the titration three times to get that level of agreement.

Volume of potassium hydroxide solution = 25.00 cm³

Burette reading	1st experiment	2nd experiment	3rd experiment
2nd reading (cm³)	17.50	19.50	20.50
1st reading (cm³)	II 0.00	2.50	3.50
Difference (cm³)	17.50	17.00	17.00

△ Table 13.1 Readings from titration.

EXAMPLE 4

In an experiment to measure the volume of a gas produced in a reaction the student has sensibly recorded her results in Table 13.2. Notice each column has a heading *and* units.

Time (s)	Volume of gas (cm³)
0	25
10	45
20	60
30	70
40	74
50	76
60	76

△ Table 13.2 Results of experiment.

4. INTERPRET OBSERVATIONS AND DATA

Learning objective: to analyse and interpret data to draw conclusions from experimental activities that are consistent with the evidence, using scientific knowledge and understanding, and to communicate these findings using appropriate specialist vocabulary, relevant calculations and graphs.

Questions to ask:

What is the best way to show the pattern in my results? Should I use a bar chart, line graph or scatter graph?

✓ Graphs are usually the best way of demonstrating trends in data.

✓ A bar chart or bar graph is used when one of the variables is a **categorical variable**; for example, when the melting points of the Group 1 elements are shown for each element, the names are categorical and not continuous variables.

✓ A line graph is used when both variables are continuous, for example, time and temperature, time and volume.

✓ Scatter graphs can be used to show the degree of *correlation* between two variables.

✓ Sometimes a line of best fit is added to a scatter graph, but usually the points are left without a line.

When drawing bar charts or line graphs:

✓ Choose scales that take up most of the graph paper.

✓ Make sure that the axes are linear and allow points to be plotted accurately. Each square on an axis should represent the same quantity. For example, one big square represents 5 or 10 units; not 3 units.

✓ Label the axes with the variables (ideally with the independent variable on the *x*-axis).

✓ Make sure the axes have units.

✓ If more than one set of data is plotted use a key to distinguish the different data sets.

If I use a line graph should I join the points with a straight line or a smooth curve?

✓ In biology, if following the biological rhythms of an organism over a period of time, you should join the data points, point-to-point.

✓ A best-fit line, straight or curved, should be drawn if a trend can be seen and there is good reason to believe that intermediate values can be predicted.

✓ A best-fit line or curve should show the 'average' trend of your plotted points. It actually does not need to go exactly through any of your plotted points – but if there are many points not exactly on the line, there should be a roughly equal distribution of them either side of the line.

✓ Remember there may be some points that don't fall on the curve – these may be incorrect or anomalous results. Any points that are clearly anomalous should be ignored when drawing your line.

✓ A graph will often make it obvious which results are anomalous.

Do I have to calculate anything from my results?

✓ It is usual to calculate means from the data.

✓ Sometimes it is helpful make other calculations, before plotting a graph; for example, you might calculate 1/time for a rate of reaction experiment in chemistry or the energy content of food *per gram* when burning a sample of food in biology.

✓ Sometimes you will have to make some calculations before you can draw any conclusions.

✓ Investigators also look for numerical trends in data, for example, the doubling of a reaction rate every 10 °C; the doubling of numbers of microorganisms every 20 minutes.

Can I draw a conclusion from my analysis of the results, and what scientific knowledge and understanding can be used to explain the conclusion?

✓ You need to use your scientific knowledge and understanding to explain your conclusion.

✓ It is important to be able to add some explanation which refers to relevant scientific ideas in order to justify your conclusion.

EXAMPLE 5

Should I join the data points point-to-point, or draw a smooth line or curve of best fit?

A student carried out an experiment to find out how the rate of a reaction changes during the reaction. She added some hydrochloric acid to marble chips and measured the volume of carbon dioxide produced in a gas syringe. She took a reading of the volume of gas in the syringe every 10 seconds for 1.5 minutes.

The apparatus she used and the results obtained are shown in Figs 13.2 and 13.3:

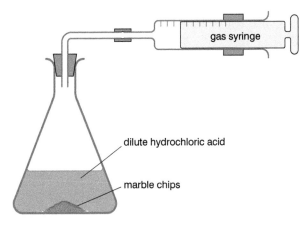

△ Fig. 13.2

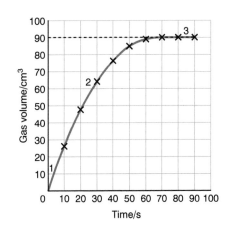

△ Fig. 13.3 Graph of experimental results.

✓ In this experiment both the volume of gas and time are continuous variables and so a line graph is needed. With the results obtained in this experiment it is clear that a smooth curve is needed. Drawing a straight line of best fit would not be sensible

Do I have to calculate anything from my results, to draw a conclusion?

The student could describe how the rate of the reaction changed as the reaction proceeded by looking at the change in steepness/gradient of the curve as the reaction proceeded, so she doesn't need to do any separate calculations.

The gradient of the line is steeper at the beginning of the experiment than nearer the end the rate of the reaction decreases as the reaction proceeds.

5. EVALUATE METHODS

Learning objective: to evaluate data and methods.

Questions to ask:

Do any of my results stand out as being inaccurate or anomalous?

✓ You need to look for any anomalous results or outliers that do not fit the pattern.

✓ You can often pick this out from a results table (or a graph if all the data points have been plotted, as well as the mean, to show the range of data).

What reasons can I give for any inaccurate results?

✓ When answering questions like this it is important to be specific. Answers such as 'experimental error' will not score any marks.

✓ It is often possible to look at the practical technique and suggest explanations for anomalous results.

✓ When you carry out the experiment you will have a better idea of which possible sources of error are more likely.

✓ Try to give a specific source of error and avoid statements such as 'the measurements must have been wrong'.

Your conclusion will be based on your findings, but must take into consideration any uncertainty in these introduced by any possible sources of error. You should discuss where these have come from in your evaluation.

Error is a difference between a measurement you make, and its true value.

The two types of error are:

✓ random error

✓ systematic error.

With **random error**, measurements vary in an unpredictable way. This can occur when the instrument you're using to measure lacks sufficient precision to indicate differences in readings. Random errors can also occur when it is difficult to make a measurement.

With **systematic error**, readings vary in a controlled way. They are either consistently too high or too low. One reason could be the way you are making a reading, for example, measuring with a measuring cylinder at the wrong point on the meniscus, or not being directly in front of an instrument when reading from it.

What an investigator *should not* discuss in an evaluation are problems introduced by using faulty equipment, or by using the equipment inappropriately. These errors can, or could have been, eliminated, by:

✓ checking equipment

✓ practising techniques before the investigation, and taking care and patience when carrying out the practical.

Overall was the method or technique I used good enough?

✓ If your results were good enough to provide a confident answer to the problem you were investigating, the method probably was good enough.

✓ If you realise your results are not accurate when you compare your conclusion with the standard result, it may be that you have a **systematic error** (an error that has been made in obtaining all the results). A systematic error would indicate an overall problem with the experimental method.

✓ If your results do not show a convincing pattern, it is fair to assume that your method or technique was not precise enough and there may have been a **random error** (that is, measurements vary in an unpredictable way).

If I were to repeat the investigation what would I change or improve?

✓ Having identified possible errors, it is important to say how these could be overcome. Again you should try and be absolutely precise.

✓ When suggesting improvements, do not just say 'do it more accurately next time' or 'measure the volumes more accurately next time'.

✓ For example, if you were measuring small lengths, you could improve the method by using a vernier scale on vernier calipers or a micrometer screw gauge to measure the lengths rather than a ruler.

✓ Investigations can also often be improved by extending the range (temperature, time, pH, and so on) over which it is carried out.

EXAMPLE 6

A student was measuring the height of a precipitate produced in a test-tube when different volumes of lead(II) nitrate solution were added to separate 15 cm³ samples of potassium iodide solution:

$Pb(NO_3)_2(aq) + 2KI(aq) \rightarrow PbI_2(s) + 2KNO_3(aq)$

Do any of my results stand out as being inaccurate or anomalous?

The student plotted her results on a graph. The inaccurate result stands out from the rest. Given the pattern obtained with the other results there is no real need to repeat the result – you could be very confident that the height should have been 3.0 cm. A result like this is referred to as an anomalous result. It was an error but not a systematic error.

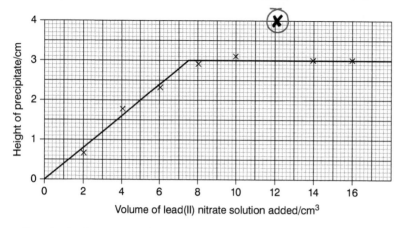

△ Fig. 13.4 Student's graph of experimental results.

What reasons can I give for any inaccurate results?

There are two main possible sources of error – either one of the volumes was measured incorrectly or the height of the precipitate was measured incorrectly.

Perhaps there was some air trapped in the precipitate and it didn't settle like the precipitates in the other tubes.

Was the method or technique I used precise enough?

You can be reasonably confident that 7.5 cm³ of lead(II) nitrate solution reacted exactly with 15 cm³ of potassium iodide solution (the point at which the height of precipitate reached its maximum value of 3.0 cm).

How can I improve the investigation?

For example, you could say, 'stir each precipitate to get rid of air bubbles and then let the precipitate settle'.

Preparing for assessment

INTRODUCTION

Examinations will test how good your understanding of scientific ideas is, how well you can apply your understanding to new situations and how well you can analyse and interpret information you have been given. The assessments are opportunities to show how well you can do these tasks.

You will need to:

✓ have a good knowledge and understanding of science

✓ be able to apply this knowledge and understanding to familiar and new situations

✓ be able to interpret and evaluate evidence that you have just been given.

You need to be able to do these things under exam conditions.

EXAMINATION TECHNIQUES

To help you to work to your best abilities in exams, there are a few simple steps to follow.

Check your understanding of the question

✓ **Read the introduction to each question carefully before moving on to the questions themselves**.

✓ Look in detail at any **diagrams, graphs** or **tables**.

✓ Underline or circle the **key words** in the question.

✓ **Make sure you answer the question that is being asked** rather than the one you wish had been asked!

✓ Make sure that you understand the meaning of the **'command words'** in the questions.

REMEMBER

Remember that any information you are given is there to help you to answer the question.

EXAMPLES

✓ **Analyse:** examine in detail to show meaning, identify elements and the relationship between them

✓ **Calculate:** work out from given facts, figures or information

✓ **Comment:** give an informed opinion

✓ **Compare:** identify/comment on similarities and/or differences

✓ **Consider:** review and respond to given information

✓ **Contrast:** identify/comment on differences

✓ **Deduce:** conclude from available information

- ✓ **Define:** give precise meaning
- ✓ **Demonstrate:** show how or give an example
- ✓ **Describe:** state the points of a topic / give characteristics and main features
- ✓ **Determine:** establish an answer using the information available
- ✓ **Discuss:** write about issue(s) or topic(s) in depth in a structured way
- ✓ **Evaluate:** judge or calculate the quality, importance, amount, or value of something
- ✓ **Examine:** investigate closely, in detail
- ✓ **Explain:** set out purposes or reasons / make the relationships between things clear / say why and/or how and support with relevant evidence
- ✓ **Give:** produce an answer from a given source or recall/memory
- ✓ **Identify:** name / select / recognise
- ✓ **Justify:** support a case with evidence/argument
- ✓ **Outline:** set out main points
- ✓ **Predict:** suggest what may happen based on available information
- ✓ **Show (that):** provide structured evidence that leads to a given result
- ✓ **Sketch:** make a simple freehand drawing showing the key features, taking care over proportions
- ✓ **State:** express in clear terms
- ✓ **Suggest:** apply knowledge and understanding to situations where there are a range of valid responses in order to make proposals / put forward considerations

The information in this section is taken from the Cambridge IGCSE Co-ordinated Sciences syllabus (0654) for examination from 2025. You should always refer to the appropriate syllabus document for the year of examination to confirm the details and for more information. The syllabus document is available on the Cambridge International website at www.cambridgeinternational.org.

Check the number of marks for each question

- ✓ Look at the **number of marks** allocated to each question.
- ✓ Look at the **space provided** to guide you as to the length of your answer.
- ✓ Make sure you include at least as many points in your answer as there are marks.
- ✓ Write neatly and keep within the space provided.

REMEMBER

Beware of continually writing too much because it probably means you are not really answering the questions. Do not repeat the question in your answer.

Use your time effectively

- ✓ Don't spend so long on some questions that you don't have time to finish the paper.
- ✓ If you are really stuck on a question, leave it, finish the rest of the paper and come back to it at the end.

✓ Even if you eventually have to guess at an answer, this gives you a better chance of getting it right than if you leave it blank.

Giving good quality answers

✓ In short-answer questions, don't write more than you are asked for.

✓ Present the information in a logical sequence.

✓ Don't be afraid to also use **labelled diagrams** or **flow charts** if it helps you to show your answer more clearly.

✓ In **calculations**, always show your working. Even if your final answer is incorrect, part of your attempt may be correct.

✓ Write down your answers to as many **significant figures** as are used in the numbers in the question (and no more). If the question doesn't state how many significant figures then a good rule of thumb is to quote 3 significant figures.

✓ Don't round off too early in calculations with many steps – it is always better to give too many significant figures than too few.

✓ Use the correct **units**. In some questions the units will be mentioned, for example: "Calculate the mass in grams". The units may also be given on the answer line.

✓ When you've finished your exam, **check through** your paper to make sure you've answered all the questions.

✓ Check that you haven't missed any questions at the end of the paper or turned over two pages at once and missed questions.

✓ Cover over your answers and read through the questions again and check that your answers are as good as you can make them.

REMEMBER

You will be asked questions on investigative work. It is important that you understand the methods used by scientists when carrying out investigative work.

Periodic Table of elements

The Periodic Table of Elements

Key

atomic number
atomic symbol
name
relative atomic mass

Group

I	II											III	IV	V	VI	VII	VIII
						1 **H** hydrogen 1											2 **He** helium 4
3 **Li** lithium 7	4 **Be** beryllium 9											5 **B** boron 11	6 **C** carbon 12	7 **N** nitrogen 14	8 **O** oxygen 16	9 **F** fluorine 19	10 **Ne** neon 20
11 **Na** sodium 23	12 **Mg** magnesium 24											13 **Al** aluminium 27	14 **Si** silicon 28	15 **P** phosphorus 31	16 **S** sulfur 32	17 **Cl** chlorine 35.5	18 **Ar** argon 40
19 **K** potassium 39	20 **Ca** calcium 40	21 **Sc** scandium 45	22 **Ti** titanium 48	23 **V** vanadium 51	24 **Cr** chromium 52	25 **Mn** manganese 55	26 **Fe** iron 56	27 **Co** cobalt 59	28 **Ni** nickel 59	29 **Cu** copper 64	30 **Zn** zinc 65	31 **Ga** gallium 70	32 **Ge** germanium 73	33 **As** arsenic 75	34 **Se** selenium 79	35 **Br** bromine 80	36 **Kr** krypto 84
37 **Rb** rubidium 85	38 **Sr** strontium 88	39 **Y** yttrium 89	40 **Zr** zirconium 91	41 **Nb** niobium 93	42 **Mo** molybdenum 96	43 **Tc** technetium –	44 **Ru** ruthenium 101	45 **Rh** rhodium 103	46 **Pd** palladium 106	47 **Ag** silver 108	48 **Cd** cadmium 112	49 **In** indium 115	50 **Sn** tin 119	51 **Sb** antimony 122	52 **Te** tellurium 128	53 **I** iodine 127	54 **Xe** xenon 131
55 **Cs** caesium 133	56 **Ba** barium 137	57–71 lanthanoids	72 **Hf** hafnium 178	73 **Ta** tantalum 181	74 **W** tungsten 184	75 **Re** rhenium 186	76 **Os** osmium 190	77 **Ir** iridium 192	78 **Pt** platinum 195	79 **Au** gold 197	80 **Hg** mercury 201	81 **Tl** thallium 204	82 **Pb** lead 207	83 **Bi** bismuth 209	84 **Po** polonium –	85 **At** astatine –	86 **Rn** radon –
87 **Fr** francium –	88 **Ra** radium –	89–103 actinoids	104 **Rf** rutherfordium –	105 **Db** dubnium –	106 **Sg** seaborgium –	107 **Bh** bohrium –	108 **Hs** hassium –	109 **Mt** meitnerium –	110 **Ds** darmstadtium –	111 **Rg** roentgenium –	112 **Cn** copernicium –	113 **Nh** nihonium –	114 **Fl** flerovium –	115 **Mc** moscovium –	116 **Lv** livermorium –	117 **Ts** tennessine –	118 **Og** oganesson –

lanthanoids

57 **La** lanthanum 139	58 **Ce** cerium 140	59 **Pr** praseodymium 141	60 **Nd** neodymium 144	61 **Pm** promethium –	62 **Sm** samarium 150	63 **Eu** europium 152	64 **Gd** gadolinium 157	65 **Tb** terbium 159	66 **Dy** dysprosium 163	67 **Ho** holmium 165	68 **Er** erbium 167	69 **Tm** thulium 169	70 **Yb** ytterbium 173	71 **Lu** lutetium 175

actinoids

89 **Ac** actinium –	90 **Th** thorium 232	91 **Pa** protactinium 231	92 **U** uranium 238	93 **Np** neptunium –	94 **Pu** plutonium –	95 **Am** americium –	96 **Cm** curium –	97 **Bk** berkelium –	98 **Cf** californium –	99 **Es** einsteinium –	100 **Fm** fermium –	101 **Md** mendelevium –	102 **No** nobelium –	103 **Lr** lawrencium –

The volume of one mole of any gas is 24 dm^3 at room temperature and pressure (r.t.p.).

Glossary

A

accurate A measurement that is close to its true value.

acid A substance that contains H^+ ions when dissolved in water and is a proton donor. It has a pH less than 7.

acid rain Rain water that contains dissolved acids, typically sulfuric acid and nitric acid.

acidic oxide The oxide of a non-metal.

activation energy The minimum energy that must be provided before a reaction can take place.

addition polymer A polymer that is made when molecules of a single monomer join together in large numbers.

addition reaction The reaction of an alkene and another element or compound to form a single compound.

alcohol A molecule with an OH group attached to a chain of carbon atoms.

alkali A base that is soluble in water and in aqueous solution produces OH^- ions. It has a pH greater than 7.

alkali metal A Group I element.

alkane A hydrocarbon in which the carbon atoms are bonded together by single bonds only.

alkene A hydrocarbon that contains a carbon–carbon double bond.

alloy A mixture of a metal and one or more other elements.

amide The functional group in the polymer nylon (nylon is a polyamide).

amine The functional group in amino acids that combines with an acid group when a protein is formed.

amphoteric oxide An oxide that reacts with both acids and alkalis to form salts.

anhydrous A substance containing no water – a compound, usually a salt, with no water of crystallisation.

anion A negatively charged ion, which moves to the anode during electrolysis.

anode A positively charged electrode in electrolysis.

atom The smallest particle of an element. Atoms are made of protons, electrons and neutrons.

Avogadro constant (or number) The number of particles in one mole of a substance. It is 6.02×10^{23}.

B

barrier methods These are used to prevent iron and steel from rusting. Examples include painting, greasing and coating with plastic.

base Substances that are oxides or hydroxides of metals and are proton acceptors.

basic oxide The oxide of a metal.

biodegradable Can be broken down naturally in the environment (for example, by bacteria in the soil).

boiling The change of state from liquid to gas.

boiling point The temperature of a boiling liquid – the highest temperature that the liquid can reach and the lowest temperature that the gas can reach.

burning The reaction of a substance with oxygen in a flame.

C

calorimetry A method for determining energy changes in reactions or when substances are mixed together.

carbonate A salt formed by the reaction of carbon dioxide with alkalis in solution.

catalyst A chemical that increases the rate of a reaction and is unchanged at the end of the reaction.

catalytic cracking The process by which long-chain alkanes are broken down to form more useful short-chain alkanes and alkenes, using high temperatures and a catalyst.

cathode A negatively charged electrode in electrolysis.

cation A positive ion, which moves to the cathode during electrolysis.

cell A device for turning chemical energy into electrical energy.

chemical change A change that is not easily reversed because new substances are made.

chemical formula The combination of element symbols that represents a compound or molecule.

chemical reaction A chemical change that produces new substances and which is not usually easily reversed.

chemical symbol A unique symbol that represents a particular chemical element.

chromatogram A visible record (usually a coloured chart or graph) showing the separation of a mixture using chromatography.

chromatography The process for separating dissolved solids using a solvent and filter paper (in the school laboratory).

climate change A shift in Earth's weather patterns, currently being caused by human activities.

collision theory A theory used to explain differences in the rates of reactions as a result of the frequency and energy associated with the collisions between the reacting particles.

combustion The reaction that occurs when a substance (usually a fuel) burns in oxygen.

complete combustion A reaction where there is sufficient oxygen to ensure that an element, or elements in a compound, are oxidised as much as they can be.

compound A pure substance formed when elements react together.

concentration Amount of chemical dissolved in 1 dm^3 of solvent (units: g/dm^3 or mol/dm^3).

condensing The change of state from gas to liquid.

condensation polymer A polymer formed when two monomers react together and eliminate a small molecule such as water or hydrogen chloride.

conductor A material that will allow heat or electrical energy to pass through it.

control variable Something that is fixed and is unchanged in an investigation.

corrosion The process in which a metal, such as iron, reacts with the oxygen in the air and water.

corrosion The process in which a metal, such as iron, reacts with the oxygen in the air and water.

covalent bond A bond that forms when a pair of electrons is shared between two atoms leading to noble gas electronic configurations.

cracking Forming shorter alkanes and alkenes from longer alkanes using high temperatures and a catalyst.

crude oil A type of fossil fuel. It is a naturally occurring liquid found below some parts of Earth's surface. It can be refined to produce fractions that have different uses.

crystallisation The formation of crystals which can occur when a solution cools or evaporates.

D

delocalised electrons Electrons that are not attached to a particular atom (as in graphite or a metallic structure).

dependent variable A variable that changes as a result of changes made to value of the independent variable.

desalination The separation of salt from sea water by evaporation of the water.

diatomic Two atoms combined together (for example in a molecule).

diffusion The random mixing and moving of particles in liquids and gases.

displacement reaction A reaction in which one element takes the place of another in a compound, removing (displacing) it from the compound.

displayed formula A displayed formula shows all the bonds in a molecule.

distillation The process for separating a liquid from a solid (usually when the solid is dissolved in the liquid) or a liquid from a mixture of liquids.

dot-and-cross diagram A diagram showing the electrons that are shared in a covalent bond between two elements. Dots are used for the electrons of one element and crosses for the electrons of the other element.

E

effective collision A collision between particles with enough energy to cause a chemical reaction.

electrical conductivity The passage of electricity through a material.

electrode The carbon or metal material that delivers electric charge in electrolysis reactions.

electrolysis The decomposition of an ionic compound, when molten or in aqueous solution, by the passage of an electric current.

electrolyte A substance that allows electric current to pass through it when it is molten or dissolved in water.

electrostatic attraction The attraction between positive and negative charges.

electron Negatively charged sub-atomic particle with a negligible mass that forms the outer part of all atoms.

electronic configuration The arrangement of electrons in an atom, molecule or ion.

element A substance that cannot be broken down into other substances by any chemical change.

end point The point in a titration when the alkali and acid are both neutralised.

endothermic A type of reaction in which thermal energy is taken in from the surroundings, leading to a decrease in the temperature of the surroundings.

enthalpy change (ΔH) The transfer of thermal energy during a reaction.

evaporation When a liquid changes to a gas at a temperature lower than its boiling point.

exothermic A type of reaction in which thermal energy is transferred to the surroundings, leading to an increase in temperature of the surroundings.

F

filtrate The liquid or solution that has passed through a filter.

filtration This is the process of separating an insoluble solid from a liquid by passing the mixture through a filter.

formula mass (M_r) The sum of the atomic masses of the atoms in a formula.

fossil fuel A fuel made from the remains of decayed animal and plant matter compressed over millions of years.

fraction A collection of hydrocarbons that have similar molecular masses and boil at similar temperatures.

fractional distillation A process for separating liquids with different boiling points.

freezing Changing a liquid to a solid.

fuel cell A cell that produces electricity from the chemical energy of a fuel. A hydrogen–oxygen fuel cell uses hydrogen and oxygen to produce electricity.

G

galvanising The process of coating a metal (usually iron) with zinc.

gas The state of matter in which the substance has no set volume or shape.

giant covalent structure A structure made of very many atoms joined together by covalent bonds.

giant lattice A giant structure of ions.

global warming The rise in the average temperature of the Earth's atmosphere and oceans.

greenhouse effect The trapping of long-wave radiation emitted from the Earth's surface by gases in the atmosphere.

greenhouse gas A gas that can trap long-wave radiation emitted from the Earth's surface.

group A vertical column of elements in the Periodic Table.

H

halogens The Group VII elements (F, Cl, Br, I, At).

homologous series A group of organic compounds with the same general formula, similar chemical properties and physical properties that change gradually from one member of the series to the next.

hydrated A substance that is chemically combined with water. Hydrated salts contain water of crystallisation.

hydrocarbon A compound containing hydrogen and carbon only.

I

impure An impure substance contains a mixture of different substances.

incomplete combustion A reaction where there is insufficient oxygen to ensure that an element, or elements in a compound, are oxidised as much as they can be.

independent variable A variable that is deliberately changed in an investigation and, as a result, causes changes to the dependent variable.

indicator A substance that changes colour in either an acid or an alkali and so can be used to identify acids or alkalis.

insoluble Does not dissolve in water.

intermolecular force The force of attraction or repulsion between molecules.

intramolecular bond A bond within a molecule.

ion A charged atom or molecule.

ionic bond A bond that involves the transfer of electrons to produce electrically charged ions.

ionic compound A compound formed by the reaction between a metal and one or more non-metals.

ionic equation A chemical equation showing how the ions involved react together.

isotopes Different atoms of the same element that have the same number of protons but different numbers of neutrons.

K

kinetic particle theory The theory describing the movement of particles in solids, liquids and gases.

L

liquid The state of matter in which a substance has a fixed volume but no definite shape.

litmus An indicator that has different colours in acids (red) and alkalis (blue).

M

malleability The measure of how easily a substance can be beaten into sheets.

mass number (nucleon number) The number of protons and neutrons in an atom (also known as the nucleon number).

melting Changing a solid into a liquid at its melting point.

melting point The temperature at which a solid changes to a liquid.

meniscus The curved upper surface of a liquid in a tube.

metal An element with particular properties (usually hard, shiny and a good conductor of heat and electricity).

metallic bonding Metallic bonding is the force of attraction between delocalised (free moving) electrons and positive metal ions.

mineral A solid inorganic substance that occurs naturally.

mixture Two or more substances combined without a chemical reaction – they can be separated easily.

molar gas volume The volume occupied, usually at standard temperature and pressure or room temperature and pressure, of one mole of any gas.

molar mass The mass of a substance in grams divided by the number of moles of the substance.

mole The amount of a substance containing 6.02×10^{23} particles (atoms, molecules, ions).

molecular formula The number and type of different atoms in one molecule.

molecule A group of two or more atoms covalently bonded together.

monatomic An element composed of separate atoms.

monomer Small molecules that can be joined in a chain to make a polymer.

N

neutralisation A reaction in which an acid reacts with a base or alkali to form a salt and water.

neutron Sub-atomic particle in the nucleus of atoms that has mass but no charge.

noble gas Group VIII (or Group 0) elements (He, Ne, Ar, Kr, Xe, Rn). They have full outer electron shells.

non-metal An element with particular properties (usually a gas or soft solid and a poor conductor of heat and electricity).

non-renewable A fuel that cannot be made again in a short time span.

nucleus The tiny centre of an atom, typically made up of protons and neutrons.

nucleon number (mass number) The total number of protons and neutrons in an atom.

— O —

ore A mineral from which a metal may be extracted.

organic chemistry The study of covalent compounds of carbon.

organic molecules Carbon-based molecules.

oxidation The gain of oxygen, loss of electrons or increase in oxidation number.

oxidation number The degree of oxidation of an element.

oxide A product of the reaction of oxygen with another element. For example, oxygen reacts with copper to produce copper oxide.

oxidising agent A substance that will oxidise another substance.

— P —

particle theory The particle theory, more often called the kinetic particle theory, states that all matter is made of very small particles that are in constant motion.

particulates Very small solid or liquid particles that are suspended in air.

period A row in the Periodic Table, from an alkali metal to a noble gas.

Periodic Table The modern arrangement of the chemical elements in groups and periods.

petroleum A mixture of hydrocarbons.

pH scale A scale measuring the acidity (lower than 7) or alkalinity of a solution (higher than 7). It is a measure of the concentration of hydrogen ions in a solution.

photosynthesis The reaction between carbon dioxide and water to produce glucose in the presence of chlorophyll and using energy from light.

physical change A change in a substance that is easily reversed and does not involve the making of new chemical bonds.

plastic A synthetic material made from a wide range of organic polymers.

polymer Large molecules built up from many smaller molecules called monomers.

polyamide A polymer with repeating units linked by amide bonds.

precipitate The solid substance formed when two liquids or solutions are mixed.

precipitation A reaction in which an insoluble salt is formed by mixing two solutions.

pressure A physical characteristic of a substance caused by particles of the substance colliding with the walls of the container it is in.

products The substances that are produced in a reaction.

proton Positively charged sub-atomic particles in the nucleus of atoms.

proton number (atomic number) The number of protons in an atom.

pure A pure substance is not a mixture, and is an element or a compound.

— R —

rate of reaction How fast a reaction goes in a given interval of time.

reactants The substances taking part in a chemical reaction. They change into the products.

reaction pathway diagram Reaction pathway diagrams show the thermal energy changes in reactions by comparing the relative energy levels of reactants and products.

reactivity series A list of elements showing their relative reactivities. More reactive elements will displace less reactive ones from their compounds.

redox A reaction involving simultaneous reduction and oxidation.

reducing agent A substance that will reduce another substance.

reduction The loss of oxygen, the gain of electrons or a decrease in oxidation number.

relative atomic mass (A_r) The average mass of the isotopes of an element compared to 1/12th of the mass of an atom of ^{12}C.

relative molecular mass (M_r) The sum of the relative atomic masses of the atoms in a molecule.

repeat unit This is the structure of the part of a polymer which when linked together shows the full polymer structure.

renewable energy Energy from a source that will not run out, such as wind, water or solar energy.

residue A substance that remains after evaporation, distillation, filtration or any similar process.

R_f value The distance travelled by a substance in a chromatography experiment (through the stationary phase) compared to the distance travelled by the solvent in the same time (expressed as a number between 0 and 1).

rusting The chemical reaction in which iron is oxidised to iron(III) oxide in the presence of air (oxygen) and water.

S

sacrificial protection Covering a metal, or ensuring contact, with another metal so that the more reactive metal corrodes instead of the less reactive metal.

salt A compound formed when the replaceable hydrogen atom(s) of an acid is (are) replaced by a metal.

saturated hydrocarbon Describes an organic compound that has molecules in which all carbon–carbon bonds are single bonds.

saturated solution A solution containing the maximum concentration of a solute dissolved in the solvent.

shell A grouping of electrons around a nucleus. The first shell in an atom can hold up to 2 electrons, the next can hold up to 8.

solid The state of matter in which a substance has a fixed volume and a definite shape.

soluble A substance that is able to dissolve in a solvent to form a solution.

solute A substance that is dissolved in a solvent.

solution A liquid mixture composed of two or more substances.

solvent A substance that dissolves a solute.

spectator ions Ions that play no part and are unchanged in a chemical reaction.

state symbols These denote whether a substance is a solid (s), liquid (l), gas (g) or is dissolved in aqueous solution (aq).

structural formula This shows how the various atoms in a structure are bonded together (it does not show all the bonds as in a displayed formula).

surface area The total area of the outside of an object.

T

thermal conductor This is a substance which is a good conductor of heat.

titration An accurate method for calculating the concentration of an acid or alkali solution in a neutralisation reaction.

transition metal Elements found between Groups II and III in the Periodic Table. Often used as catalysts and often make compounds that have coloured solutions.

trend This can be used to describe the change in the physical or chemical properties of a series of substances (e.g. elements).

U

universal indicator Indicator solution that turns a specific colour at each pH value.

unsaturated hydrocarbon Describes molecules in which one or more of the carbon–carbon bonds are double bonds or triple bonds.

V

vapour Another term for gas.

variable A factor that can either be changed in an investigation or changes as a result of other factors changing.

volatile Easily turns to a gas.

Answers

All answers, including answers to practice questions, have been written by the authors. In examinations, the way marks are awarded may be different.

These are the answers to the questions in each topic. Answers to end-of-topic and practice questions are available on the Collins website.

SECTION 1 STATES OF MATTER

States of matter

Page 11

1. (l)
2. Only the solid state has a fixed shape.
3. Fine sand will pour or flow like a liquid; it takes the shape of the container it is poured into (although under a microscope you would see gaps at the surface of the container).

Page 15–16

1. The particles in a solid vibrate about a fixed point.
2. The particles are held together the most strongly in solid water (ice).
3. Evaporation is the process that occurs when faster-moving particles in a liquid escape from the liquid surface.
4. Melting is the name of the process when a solid changes into a liquid.
5. When the pressure of a gas is increased the particles are forced closer together taking up less space and so reducing the volume of the gas.
6. A and C

Page 18

1. Diffusion is the mixing and moving of particles in liquids and gases.
2. The particles in the potassium manganate(VII) dissolve in the water and diffuse throughout the solution.

3. The particles of perfume vapour/gas diffuse in the air and spread throughout the whole room.
4. Hydrogen has a lower relative molecular mass than oxygen so the particles of hydrogen will move more rapidly at a particular temperature than those of oxygen. As a result the hydrogen will diffuse more rapidly than oxygen.

SECTION 2 ATOMS, ELEMENTS AND COMPOUNDS

Materials, atomic structure and the Periodic Table

Page 27

1. The electron has the smallest relative mass.
2. Atoms are neutral. The number of positive charges (protons) must equal the number of negative charges (electrons).
3. **a)** The mass number is 27.
   b) 14 neutrons

Page 28

1. Isotopes are atoms of the same element with different numbers of neutrons.
2. Isotopes of the same element have the same number of electrons. It is the number of electrons which determine the chemical properties of the element.

Page 32

1. **a)** Magnesium has two electrons in its outer electron shell.
   b) It is in Group II.

2. a) Aluminium

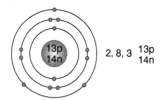

2, 8, 3 13p 14n

b) Calcium

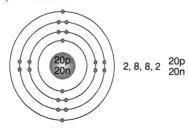

2, 8, 8, 2 20p 20n

Ions and ionic bonds

Page 39

1. A cation is a positive ion.

2.

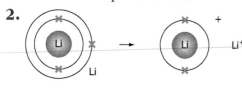

3.

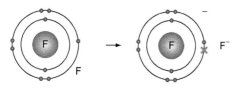

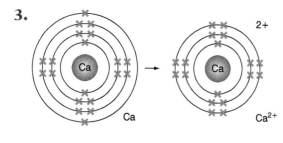

4. Both phosphorus and oxygen are non-metals. (A metal is needed to form an ionic bond.)

Page 42

1. High melting point, high boiling point, good conductor of electricity when aqueous or molten.

2. The ions are held together strongly in a giant lattice structure. The ions can vibrate but cannot move around.

3. Sodium chloride is made up of singly charged ions, Na^+ and Cl^-, whereas the magnesium ion in magnesium oxide has a double charge, Mg^{2+}. The higher the charge on the positive ion, the stronger the attractive forces between the positive ion and the negative ion.

Molecules and covalent bonds

Page 49

1. Both hydrogen atoms have 2 electrons in their outer electron shells (like helium).

2. Each chlorine atom has 17 electrons.

3. The nitrogen atom has 5 electrons in its outermost electron shell and needs an additional 3 electrons to achieve the inert gas structure of a full outer shell. Each hydrogen atom can only provide 1 electron.

4. Each oxygen atom needs 2 more electrons to complete their outer electron shells. A double covalent bond provides 2 additional electrons to each oxygen atom.

5.

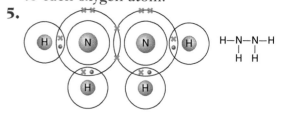

Page 51

1. The intermolecular forces of attraction between the molecules are weak.

2. No. There are no ions or delocalised electrons present.

ANSWERS

Page 53

1. Each carbon atom is strongly covalently bonded to four other atoms forming a very strong giant lattice structure. A very high temperature is needed to break down the structure.

2. In graphite each carbon atom is strongly covalently bonded to three other carbon atoms. The remaining outer shell carbon electron is delocalised and so can move along between the layers formed by the covalently bonded carbon atoms.

Metallic bonding

Page 57

1. A cation is a positive ion.
2. Metals contain delocalised electrons that are not fixed to a particular atom; they can move throughout the structure.
3. Metals are not rigid, so the ions can move into different positions when the metal is bent.

SECTION 3 STOICHIOMETRY

Formulas

Page 67–68

1. HCl – 1 hydrogen atom and 1 Chlorine atom

   F_2 – 2 fluorine atoms

   CH_3 – 1 carbon atom, 3 hydogen atoms

   H_2O – 2 hydrogen atoms and 1 oxygen atom

   LiOH – 1 lithium atom, 1 oxygen atom, 1 hydrogen atom

   $MgCl_2$ – 1 magnesium atom, 2 chlorine atoms

   $CaCO_3$ – 1 calcium atom, 1 carbon atom, 3 oxygen atoms.

2. a) KBr
   b) CaO
   c) $AlCl_3$
   d) CH_4

3. a) C_4H_{10}
   b) C_3H_6

4. a) $ZnCl_2$
   b) Cr_2O_3
   c) $Fe(OH)_2$

5. NaCl

Page 70

1. a) $2Ca + O_2(g) \rightarrow 2CaO(s)$
   b) $2H_2S(g) + 3O_2(g) \rightarrow 2SO_2(g) + 2H_2O(l)$

2. a) $C(s) + O_2(g) = CO_2(g)$
   b) $2Mg(s) + O_2(g) \rightarrow 2MgO(s)$
   c) $CuO(s) + H_2(g) \rightarrow Cu(s) + H_2O(l)$

Page 71

1. a) $2C_5H_{10}(g) + 15O_2(g) \rightarrow 10CO_2(g) + 10H_2O(l)$
   b) $Fe_2O_3(s) + 3CO(g) \rightarrow 2Fe(s) + 3CO_2(g)$
   c) $2KMnO_4(s) + 16HCl(aq) \rightarrow 2KCl(s) + 2MnCl_2(s) + 8H_2O(l) + 5Cl_2(g)$

Page 74

1. 16
2. 46
3. 48

Page 76

1. a) To allow oxygen from the air into the crucible.
   b) To limit the loss of magnesium oxide.
   c) White.

Page 77

1. Fe_2O_3
2. ZnO

Page 80

1. 28 g
2. 10 g

Page 80–81

1. **a)** 2 moles
   b) 0.5 mole
   c) 0.1 mole
2. **a)** 0.5 mole
   b) 0.1 mole
   c) 2 moles

SECTION 4 ELECTROCHEMISTRY

Electrochemistry

Page 98

1. The breaking down (decomposition) of an ionic compound by the use of electricity.
2. The positive electrode is the anode.
3. The substance must contain ions and they must be free to move (in molten/ liquid state or dissolved in water).

Page 101

1. **a)** An inert electrode is an unreactive electrode; it will not be changed during electrolysis.
   b) Carbon is commonly used as an inert electrode (Platinum is another inert electrode.)
2. **a)** Lead and chlorine.
   b) Magnesium and oxygen.
   c) Aluminium and oxygen.
3. **a)** $Al^{3+} + 3e^- = Al$
   b) The change takes place at the anode.

SECTION 5 CHEMICAL ENERGETICS

Chemical energetics

Page 111

1. A reaction that releases energy to the surroundings.
2. A reaction that absorbs energy from the surroundings.
3. Very little energy is transferred to the surroundings.

Page 112

1. A high proportion of the energy released transfers to the surrounding air.
2. **a)** Weighing the spirit burner before and after burning the fuel.
   b) Ethanol 29; Paraffin 33; Pentane 25; Octane 40.
   c) i) Octane
   ii) A 10 °C rise would be expected. The same amount of energy is transferred to double the volume of water.
3. The group of students using the metal should get more accurate results. The metal can conducts the heat from the fuel to the water better than glass does.

Page 114

1. The reaction is endothermic.
2. In an endothermic reaction, the energy released when the products are formed is less than the activation energy (the energy absorbed by the reactants.)
3. The activation energy.
4.

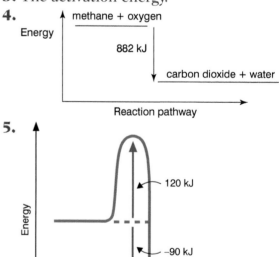

5.

Reaction A

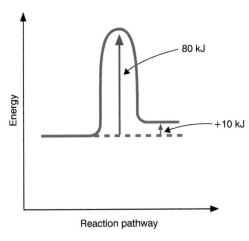

Energy

80 kJ

+10 kJ

Reaction pathway

Reaction B

Page 116

1. The sign indicates whether the reaction is exothermic (negative sign) or endothermic (positive sign).
2. Energy is needed to break bonds.
3. In an endothermic reaction more energy is needed to break bonds than is recovered on forming bonds.

SECTION 6 CHEMICAL REACTIONS

Rate of reaction

Page 126

1. In a physical change no new substances are made. In a chemical change at least one new substance is made.
2. The apparent change in mass is often because a gas has been either a reactant or a product (and is lost from the reaction vessel).
3. The particles must collide; there must be sufficient energy in the collision (to break bonds).
4. An effective collision is one which results in a chemical reaction between the colliding particles.
5. Student's diagram like Fig. 6.3. It is an energy barrier. Only collisions that have enough energy to overcome this barrier will lead to a reaction.

1. A gas syringe will measure the volume of gas produced accurately.
2. No gas is being produced – the reaction hasn't started or it is finished.
3. The quicker reaction will have the steeper gradient.

Page 132

1. The particles are closer together and there are more of them, so there will be more (effective) collisions per second.
2. a) Risk assessments included, quantities of magnesium and hydrochloric acid (concentration specified), apparatus, e.g. for collection of gas (gas syringe) or loss in mass (electronic balance).
   b) Appropriate sequence of steps in the method with a sensible number of different temperatures chosen.
   c) Appropriate table of results including the units of the measurements to be taken.
   If possible, students could carry out their plan as a class practical.
3. Increasing the temperature means the particles have more (kinetic) energy. So more of the collisions will have energy greater than or equal to the activation energy and there will be more effective/successful collisions per second.

Page 134

1. A catalyst is a substance that changes the rate of a chemical reaction.
2. A reaction involving gases (at least one gas).
3. The larger the surface area of the solid the greater the number of effective collisions with the acid particles and therefore the greater the rate of reaction.

Redox reactions

1. Reduction is the loss of oxygen or the gain of electrons.
2. Aluminium + oxygen = aluminium oxide
3. Carbon – oxidised
   Copper(II) oxide – reduced.
4. Oxidised, as it has lost electrons.

SECTION 7 ACIDS, BASES AND SALTS

Acids, bases and salts

Page 148

1. Both solutions are alkalis. Solution A is weakly alkaline whereas solution B is strongly alkaline.
2. The solution is acidic.
3. Calcium is a metal. The oxides (and hydroxides) of metals are bases.

Page 150

1. Potassium oxide is a basic oxide. Potassium is a metal and most metal oxides are basic.
2. A basic oxide reacts with acids and not alkalis. An amphoteric oxide reacts with both acids and alkalis (to form salts).
3. $4K(s) + O_2(g) = 2K_2O(s)$

Page 152

1. a) magnesium + sulfuric acid $\rightarrow$ magnesium sulfate + hydrogen
   b) copper(II) oxide + hydrochloric acid $\rightarrow$ copper(II) chloride + water
   c) copper(II) carbonate + nitric acid $\rightarrow$ copper(II) nitrate + carbon dioxide + water
2. a) $Mg + 2H^+ \rightarrow Mg^{2+} + H_2$
   b) $CO_3{}^{2-} + 2H+ \rightarrow CO_2 + H_2O$

Page 154

1. A salt is formed when a replaceable hydrogen of an acid is replaced by a metal.

2. Sulfuric acid
3. Calcium nitrate
4. Neutralisation is the reaction between and acid and an alkali or base to form a salt and water.

Pages 155

1. A precipitation reaction is the formation of an insoluble salt as a result of a chemical reaction taking place in aqueous solution.
2. Filtration
3. Washing with cold water will remove traces of any remaining soluble salts.
4. a) lead(II) nitrate + sodium chloride $\rightarrow$ lead(II) chloride + sodium nitrate
   b) $Pb(NO_3)_2(aq) + 2NaCl(aq) \rightarrow PbCl_2(s) + 2NaNO_3(aq)$

SECTION 8 THE PERIODIC TABLE

The Periodic Table

Page 170

1. a) 20
   b) The proton number is the number of protons (which equals the number of electrons) in an atom of the element. Calcium atoms have 20 protons and 20 electrons.
   c) Group II
   d) Period 4
   e) Calcium is a metal.
2. The halogens
3. The halogens are non-metals.

Page 171

1. Aluminium has 3 electrons in the outer shell.
2. Elements in Group VI are likely to be non-metals as there is a trend from metallic to non-metallic across a period of the Periodic Table.
3. Flourine (F)

Group I elements

1. They react with water to form alkaline solutions.
2. They have one electron in the outer shell.
3. They are soft to cut (also have very low melting points).
4. The melting point of potassium will be lower than that for sodium as melting points decrease down Group I.
5. a) 40 ±5°C (The actual m.p. is 39.5°).
   b) Very fast (high rate).

1. Hydrogen. The solution formed is potassium hydroxide.
2. a) A period is a horizontal row of elements in the Periodic Table. All the elements in the same period have the same number of electron shells.
   For example: Period 1 elements: hydrogen (1) to helium (2) – 1 electron shell.
   Period 2 elements: lithium (2,1) to neon (2,8) – 2 electron shells.
   Period 3 elements: sodium (2,8,1) to argon (2,8,8) – 3 electron shells.
   b) Magnesium is in the 3rd period of the Periodic Table.
   c) Sodium has only 1 electron in its outer shell and this is relatively easily removed as it is shielded from the nucleus by the other shells. Argon has a full shell of electrons and so is very unreactive.
3. a) A group is a vertical column of elements having similar chemical properties because of their outer shell electronic structure.
   b) lithium, sodium, potassium
   c) All the elements in Group I have one electron in the outer shell.

Group VII elements

1. Seven electrons in the outer shell.
2. The atoms need to gain only one electron to achieve a full outer shell.
3. Chlorine molecules are made up of two atoms combined/ bonded together, Cl_2.
4. Astatine is probably a solid at room temperature. Group VII elements show trends in their physical and chemical properties. At room temperature, fluorine and chlorine are gases, bromine is a liquid and iodine is a solid. Astatine is below iodine in Group VII, therefore it is also likely to be a solid.
5. A displacement reaction is one where one halogen element takes the place of another in a compound.
6. $Br_2(l) + 2I^-(aq) = 2Br^-(aq) + I_2(s)$

1. a)

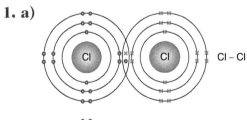

   b) $2Cl^-(aq) \rightarrow Cl_2(g) + 2e^-$. The chloride ions lose electrons.
   c) Chlorine is a more reactive halogen than bromine. Chlorine will displace bromine from a solution of a bromide ions. (Chlorine will oxidise bromide ions to bromine.)

Observations: chlorine water is pale green. When this is added to a colourless solution of potassium bromide the resulting solution will turn orange due to the presence of bromine.

2. a) $F_2 + 2e^- \rightarrow 2F^-$
   b) Both chlorine and iodine are less reactive than fluorine. Fluorine could only be displaced from fluoride ions by a more reactive halogen. As there are no halogens that are more reactive than fluorine, fluorine will not be displaced from fluoride ions.

Transition metals and noble gases

Page 190–191

1. a) 30
   b) The number of protons which equals the number of electrons.
   c) 4th period
   d) Metal
2. No. Copper is very unreactive – it is below hydrogen in reactivity series.
3. The number indicates the oxidation number of the chromium.
4. a) $FeSO_4(aq) + 2NaOH(aq) \rightarrow$
      $Fe(OH)_2(s) + Na_2SO_4(aq)$
   b) Green

Page 191 (bottom)

1. It has 8 electrons in its outer shell.
2. It exists as single atoms.
3. a) 18.
   b) 3, so 3 shells.
   c) 8.
   d) 2, 8, 8.

SECTION 9 METALS

Metals

Page 200

1. A malleable metal can be hammered into shape.

2. An alloy is a mixture of a metal with one or more other elements.
3. The element added in the alloy disrupts the rows of aluminium atoms making them less likely to slide over each other when under strain.

Page 202–03

1. No. Copper is below hydrogen in the reactivity series.
2. a) Risk assessments included, quantities of metals and hydrochloric acid (concentration specified) and apparatus.
   b) Appropriate sequence of steps in the method.
   c) Appropriate table of results including the units of the measurements to be taken.
   If possible, students could carry out their plan as a class practical.
3. $2K(s) + 2H_2O(l) \rightarrow$
   $2KOH(aq) + H_2(g)$
4. No. Carbon is below magnesium in the reactivity series.
5. $Mg(s) + Pb^{2+}(s) \rightarrow Mg^{2+}(s) + Pb(s)$

Page 205

1. Air (oxygen) and water must be present.
2. The grease can be easily removed or wiped away.
3. a) Galvanising involves coating iron or steel with zinc.
   b) Because zinc is more reactive than iron, moist air will react with zinc in preference to the iron.

Pages 207

1. Iron ore (hematite), coke and limestone.
2. Iron(III) oxide.
3. Carbon dioxide, carbon monoxide, nitrogen (from the air).

4. $2Fe_2O_3(s) + 3C(s) \rightarrow$
$\qquad\qquad 4Fe(s) + 3CO_2(g)$
5. a) An alloy is a mixture of a metal and another element.
   b) Steel is stronger and generally more resistant to corrosion.
6. a) oxygen + water
   b) Coating iron (or steel) with zinc provides a barrier preventing oxygen and water reaching the iron. In addition, as zinc is more reactive than iron it will corrode more rapidly than iron so protecting the iron even further.

SECTION 10 CHEMISTRY OF THE ENVIRONMENT

Chemistry of the environment

Page 220

1. Anhydrous means without water (water of crystallisation).
2. Cobalt(II) chloride will change from blue to pink.
3. Measure the freezing point – pure water will freeze at $0\,°C$; measure the boiling point – pure water will boil at $100\,°C$ (under conditions of standard pressure).
4. a) The first filter is coarse gravel. The second filter is fine sand.
   b) Chlorine is used to kill bacteria.
5. a) Nitrogen is 78%.
   b) Carbon dioxide is 0.04%.

Page 225

1. A major source of carbon monoxide is the incomplete combustion of fuels, such as in a motor car.
2. a) Methane is the major component of natural gas. It is also produced by decaying vegetable matter and by ruminant animals such as cows.
   b) Carbon dioxide is another greenhouse gas.

3. a) Sulfur dioxide and nitrogen oxide(s) are gases that cause acid rain.
   b) Sulfur dioxide forms sulfuric acid; nitrogen oxide forms nitric acid.
   c) Environmental problems include: harming plants and fish in lakes, damaging buildings made of metal, marble or limestone.
4. a) $N_2(g) + O_2(g) \rightarrow 2NO(g)$
   b) $2NO(g) + 2CO(g) \rightarrow$
   $\qquad\qquad N_2(g) + 2CO_2(g)$

SECTION 11 ORGANIC CHEMISTRY

Fuels

Page 239

1. The supplies of petroleum are limited – it takes millions of years for crude oil to be formed.
2. The attractive forces between the molecules increase as the carbon chain length increases. Therefore, a fraction with a smaller chain length is more likely to be a gas.
3. Long chain of carbon atoms.

Alkanes

Page 245 (top)

1. a) A compound that has no C=C double bonds.
   b) A compound containing hydrogen atoms and carbon atoms only.
2. a) $C_{15}H_{32}$
   b) Carbon dioxide and water.

Page 245 (bottom)

1. The fuel will burn with a yellow (rather than blue) flame.
2. Carbon and carbon monoxide.
3. Wind, wave, solar and nuclear power are alternative ways of generating energy.

Alkenes

Page 250

1. The fractional distillation of crude oil produces a high proportion of long-chain hydrocarbons, which are not as useful as short-chain hydrocarbons. Cracking converts the long-chain hydrocarbons into more useful short-chain hydrocarbons.
2. The conditions required for cracking oil fractions are a temperature of between 600 and 700 °C and a catalyst of silica or alumina.

Page 253

1. It contains at least one C=C double bond.
2. Ethene is manufactured by the catalytic cracking of larger alkane molecules using a high temperature.
3. The large alkane molecules are not very useful whereas the alkene molecules can be used to manufacture a range of useful products, for example, polymers.
4. C_3H_6

Alcohols

Page 260

1. C_2H_5OH
2. It is a relatively 'clean' fuel, releasing carbon dioxide and water into the atmosphere. It does not release sulfur dioxide and nitrogen oxides which are the main causes of acid rain, as petrol does when it burns. Increasingly, however, the production of carbon dioxide and its impact on global warming is becoming a very big issue.
3. A solvent is a liquid that dissolves other substances (solutes) to form solutions.

Polymers

Page 268

1. Nylon is made from two monomers and a small molecule is eliminated when the two monomers combine. An addition polymer has only one monomer.
2. To produce a polymer chain the monomers need to be able to form amide groups at both ends of the molecules.

3.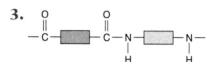

SECTION 12 EXPERIMENTAL TECHNIQUES AND CHEMICAL ANALYSIS

Experimental techniques

Page 283

1. A baseline drawn in pencil will not dissolve in the solvent.
2. If the solvent were above the baseline the substances would just dissolve and form a solution in the beaker.
3. The dye may be insoluble in the solvent.
4. The boiling point will be higher than that of pure water/ above 100 °C at normal pressure.
5. $R_f = 1.7/10 = 0.17$

Page 285

1. A solvent is a liquid that will dissolve a substance (the solute).
2. If a substance is soluble in a solvent it dissolves in that solvent.
3. Distillation
4. Boiling points

1. a) A white precipitate, which does not dissolve in excess sodium hydroxide solution.
   b) A white precipitate, which does dissolve in excess sodium hydroxide solution.
2. Add sodium hydroxide solution. Fe^{2+} produces a green precipitate; Fe^{3+} produces a reddish-brown precipitate.

Page 294–295

1. Add dilute sodium hydroxide and heat. An alkaline gas (turns red litmus paper blue) indicates the presence of an ammonium compound.
2. a) Carbon dioxide
   b) Bubble the gas through limewater. A white precipitate forms.
3. The Fe^{3+} ion is present in solution X.
4. The Cl^- ion is present in solution Y.
5. Plan needs to check for testing of both anion and cation for each sample and should include practical instructions.
   Blue compound
   - Test for copper(II) – sodium hydroxide: result blue precipitate $Cu^{2+}(aq) + 2OH^-(aq) \rightarrow Cu(OH)_2(s)$

- Test for sulfate – hydrochloric acid/ barium chloride: result white precipitate $Ba^{2+}(aq) + SO_4^{2-}(aq) \rightarrow BaSO_4(s)$
White compound
- Flame test for Na^+ – yellow
- Test for carbonate – add dilute acid – effervescence/carbon dioxide evolved – turns limewater milky $CO_3^{2-}(s) + 2H^+(aq) \rightarrow H_2O(l) + CO_2(g)$

6. a)

Name of cation	Colour of precipitate
Zinc/lead	White
Magnesium/ calcium	White
Copper(II)	Blue
Iron(II)	green/turns brown slowly
Iron(III)	rust brown/ orange

b) $Ag^+(aq) + X^-(aq) \rightarrow AgX(s)$ where X^- is Cl^-, Br^-, I^-.

Page 296

1. Ammonia is tested using damp red litmus paper – the paper turns blue.
2. Oxygen
3. Chlorine

Index

Acknowledgements

The publishers wish to thank the following for permission to reproduce photographs. Every effort has been made to trace copyright holders and to obtain their permission for the use of copyright materials. The publishers will gladly receive any information enabling them to rectify any error or omission at the first opportunity:

(t = top, c = centre, b = bottom, r = right, l = left)

Cover and title page image Ann Paganuzzi, p4 Tatiana Gasich/Shutterstock, p5tl Graham Taylor/Shutterstock, p5tr Bokeh Art Photo/Shutterstock, p5tr Rvkamalov gmail.com/Shutterstock, p5tr Giovanna Di Mauro/Shutterstock, p5bl Andrew Lambert Photography/Science Photo Library, p5bc Andrew Lambert Photography/Science Photo Library, p5br ANDREW MCCLENAGHAN/Science Photo Library, p6br Andrew Lambert Photography/Science Photo Library, pp8-9 DCrane/Shutterstock, p10t jele/Shutterstock, p10b Achim Baque/Shutterstock, p14 cobalt88/Shutterstock, p16tl Andrew Lambert Photography/Science Photo Library, p16tr Andrew Lambert Photography/Science Photo Library, p16br ANDREW MCCLENAGHAN/Science Photo Library, pp22-23 Tatiana Gasich/Shutterstock, p24 Andrew Lambert Photography/Science Photo Library, p27 PJF Military Collection/Alamy Stock Photo, p28 David Parker & Julian Baum/Science Photo Library, p32 Philip Ho/Shutterstock, p36 Smit/Shutterstock, p40 Blaz Kure/Shutterstock, p41tr agefotostock/Alamy Stock Photo, p41cl DiamondGalaxy/Shutterstock, p41cr Marc Dietrich/Shutterstock, p45 Voronin76/Shutterstock, p52t broukoid/Shutterstock, p52b Tyler Boyes/Shutterstock, p53t Dmitry Kalinovsky/Shutterstock, p53b Tatiana Grozetskaya/Shutterstock, p56 demarcomedia, p58 liseykina/Shutterstock, pp62-63 NavinTar/Shutterstock, p64 Martyn F. Chillmaid/Science Photo Library, p71 Andrew Lambert Photography/Science Photo Library, p72c E.R. Degginger/Alamy Stock Photo, p72b pics five/Shutterstock, p74 Andrew Lambert Photography/Science Photo Library, p79 ARENA Creative/Shutterstock, p82 Science Photo Library, pp94-95 HR Media Art/Shutterstock, p96 MAXIMILIAN STOCK LTD/SCIENCE PHOTO LIBRARY, p102 Friedrich Saurer/Alamy Stock Photo, p103 Keitma/Shutterstock, p104 Glenmore/Shutterstock, pp108-109 Aleksay Sagitov/Shutterstock, p110 badahos/Shutterstock, p117tl Iakov Kalinin/Shutterstock, p117br Sheila Fitzgerald/Shutterstock, pp122-123 ZHMURCHAK/Shutterstock, p124 Kaweewat/Shutterstock, p132 Martyn F. Chillmaid/Science Photo Library, p133l dgmata/Shutterstock, p133r Isolated Pro/Shutterstock, p134 CHARLES D. WINTERS/SCIENCE PHOTO LIBRARY, p138 Cyril Hou/Shutterstock, p140 Phil Degginger/Alamy Stock Photo, pp144-145 Smereka/Shutterstock, p146 MARTYN F. CHILLMAID/Science Photo Library, p149bl Andrew Lambert Photography/Science Photo Library, p149br Andrew Lambert Photography/Science Photo Library, p151l Blaz Kure/Shutterstock, p151r Andrew Lambert Photography/Science Photo Library, p153 CHARLES D. WINTERS/Science Photo Library, p155 jcwait/Shutterstock, pp166-167 DimaBerlin/Shutterstock, p168 Shebeko/Shutterstock, p174 Andrew Lambert Photography/Science Photo Library, p175 CHARLES D. WINTERS / SCIENCE PHOTO LIBRARY, p177 Andrew Lambert Photography/Science Photo Library, p178 design56/Shutterstock, p181 Andrew Lambert Photography/Science Photo Library, p189 Slaven/Shutterstock, pp194-195 John Kellerman/Alamy Stock Photo, p196 Graham Taylor/Shutterstock, p197bl Bokeh Art Photo/Shutterstock, p197bc Rvkamalov gmail.com/Shutterstock, p197br Giovanna Di Mauro/Shutterstock, p198 2xWilfinger/Shutterstock, p200tl LAWRENCE MIGDALE/SCIENCE PHOTO LIBRARY, p200tr RICHARD TREPTOW/SCIENCE PHOTO LIBRARY, p200bl Julia Reschke/Shutterstock, p204 Joe Gough/Shutterstock, p206tl Fokin Oleg/Shutterstock, p206tr kilukilu/Shutterstock, p206cr Parmna/Shutterstock, p206cl Holly Kuchera/Shutterstock, p206br Danicek/Shutterstock, p208 World History Archive/Alamy Stock Photo, pp214-215 Mike Pellinni/Shutterstock, p216 StrippedPixel/Shutterstock, p217 MARTYN F. CHILLMAID/Science Photo Library, p221 Nando Machado/Shutterstock, p224 muzsy/Shutterstock, pp234-235 curraheeshutter/Shutterstock, p236 Paul Rapson/Science Photo Library, p237 BESTWEB/Shutterstock, p239 DeZet/Shutterstock, p242 Yvan/Shutterstock, p244tr speedpix/Alamy Stock Photo, p244bl ggw/Shutterstock, p245 Joe Gough/Shutterstock, p248 Gwoeii/Shutterstock, p254 JoLin/Shutterstock, p258 Tonis Valing/Shutterstock, p260 David R. Frazier Photolibrary, Inc./Alamy, p261 Jim Parkin/Alamy, p265 Dmitry Yashkin/Shutterstock, p267 kaband/Shutterstock, p268 EpicStockMedia/Shutterstock, p269tr photka/Shutterstock, p269bl Roman Zaiets/Shutterstock, pp276-277 Gorodenkoff/Shutterstock, p278 CHARLES D. WINTERS/Science Photo Library, p279 haveseen/Shutterstock, p281l Andrew Lambert Photography/Science Photo Library, p281r Andrew Lambert Photography/Science Photo Library, p283 Martyn F. Chillmaid/Science Photo Library, p284bl Ajamal/Shutterstock, p289 Johann Helgason/Shutterstock, p290 Noiel/Shutterstock, p291 Andrew Lambert Photography/Science Photo Library, p294 Andrew Lambert Photography/Science Photo Library, p304 Ed Phillips/Shutterstock.